The Institute of Biology's
Studies in Biology no. 49

Estuarine Biology
Second edition

R. S. K. Barnes

B.Sc., M.A., Ph.D.

St. Catharine's College and Zoological Laboratory,
Cambridge

Edward Arnold

First published in Great Britain 1974
by Edward Arnold (Publishers) Ltd
41 Bedford Square, London WC1B 3DQ

Edward Arnold (Australia) Pty Ltd
80 Waverley Road
Caulfield East
Victoria 3145
Australia

Edward Arnold
300 North Charles Street
Baltimore
Maryland 21201
United States of America

Reprinted with additions and amendments 1976
Second edition 1984

British Library Cataloguing in Publication Data

Barnes, R.S.K.
Estuarine biology.—2nd ed.—(The Institute of Biology's studies in biology, ISSN 0537-9024; no. 49)
1. Estuarine ecology
I. Title II. Barnes, R.S.K. Estuarine ecosystems III. Series
574.5 '26365 QH541.5.E8

ISBN 0-7131-2905-0

Text set in 9/11pt Times by The Castlefield Press
Printed and bound in Great Britain at
The Camelot Press Ltd, Southampton

General Preface to the Series

Because it is no longer possible for one textbook to cover the whole field of biology while remaining sufficiently up to date, the Institute of Biology proposed this series so that teachers and students can learn about significant developments. The enthusiastic acceptance of 'Studies in Biology' shows that the books are providing authoritative views of biological topics.

The features of the series include the attention given to methods, the selected list of books for further reading and, wherever possible, suggestions for practical work.

Readers' comments will be welcomed by the Institute.

1984

Institute of Biology
20 Queensberry Place
London SW7 2DZ

Preface to the First Edition

This booklet attempts to view the biology of estuarine organisms in the context of their environment as a whole. It also endeavours to introduce to a wide audience some of the more recent advances in our knowledge and understanding of estuaries. The author, at least, will be more than satisfied if the booklet succeeds in providing a general framework against which some of the detailed studies on estuarine species can be considered; in stimulating interest in some of the as yet unresolved questions; and in conveying some of the pleasure and satisfaction to be derived from studying our estuaries.

Should my botanical colleagues consider that too great an emphasis has been placed upon the fauna, I can only admit that were I a botanist I would probably agree with them. To give a complete picture of the many-faceted biology of estuaries in a small booklet is clearly impossible. Rather than write one short paragraph on each facet, I have been selective, choosing what to me are the most important and interesting of recent developments, whilst still endeavouring to tell a unified story. The suggested material for further reading will hopefully permit the reader to delve into fields which pressure of space has forced me to neglect.

To all those friends and colleagues who were inveigled into reading all or part of this booklet in draft, and especially to my wife, may I offer my most sincere gratitude; through their efforts the booklet has been immeasurably improved.

Cambridge, April 1973 R.S.K.B.

Preface to the Second Edition

The last 10 years have seen a considerable advancement in our knowledge and understanding of estuarine biology, particularly in the fields of ecology and animal behaviour. This second edition has therefore provided an opportunity to rewrite much of Chapters 3 and 4, and to bring up to date the material in the other parts of the booklet. I have, however, retained the basic form and coverage of the first edition.

Cambridge, January 1984 R.S.K.B.

Contents

1 Introduction: the Environment

1.1 What is an estuary?

Most people know what is meant by 'an estuary': it is a region through which a river discharges into the sea. But, like many situations in biology, although we can easily form the appropriate mental picture, we experience difficulties when attempting to define the situation more rigidly. Indeed, there are probably almost as many formal notions of the precise nature of an estuary as there are specialist estuarine scientists.

The difficulties involved in trying to put formal limits on what is essentially a continuum can best be illustrated by consideration of two extreme cases: a large river (e.g. the Amazon) discharging directly into the sea, and a small river or rivers discharging into a large drowned valley (e.g. Southampton Water).

A glance at an atlas tells us that the mouth of the Amazon continually widens the nearer one gets to the Atlantic Ocean. We therefore have a picture suggestive of the situation in the Thames Estuary on a large scale. But, in contrast to the Thames, the volume of fresh water discharged by the Amazon is so large (i.e. almost 20% of the total river discharge into the world's oceans) that only under drought conditions does sea water penetrate into what, from geographical considerations, one might expect to be its estuary. The region of mixing between the discharged fresh water and the recipient salt water occurs in a plume out at sea (Fig. 1–1). Hence the 'true estuary' of the Amazon is not bordered by land. Although the lower reaches of the Amazon are not rendered brackish, the sea does influence the river in another respect: the Amazon is tidal for a distance of at least 450 miles upstream from its mouth due to ponding back of the fresh water at high tide.

On turning to drowned valleys such as Southampton Water, one finds that the volume of fresh water discharged is normally very small in comparison with that of the receiving water. Hence Southampton Water is virtually an arm of the English Channel, although almost entirely enclosed by land (Fig. 1–2).

What characters, then, are common to the situations that one finds associated with the Amazon and with Southampton Water? The one important feature shared by both is that a body of water of mixed origin exists, with a freshwater component supplied by the appropriate river system and a marine component derived from the adjacent sea. In most estuaries, this body of water of mixed origin does not occupy a constant position: it may move with the state of the tide and with changes in the volume of fresh water discharged. The zone of mixing is only static if the freshwater discharge is constant throughout the year and if the tidal amplitude is negligible.

This analysis would lead us to suggest, as a working definition for this book, that 'an estuary is a region containing a volume of water of mixed origin derived partly from a discharging river system and partly from the adjacent sea; the region usually being partially enclosed by a land mass'. Hence an estuary is primarily a hydrographical phenomenon.

The geomorphological character of estuaries and their modes of formation are varied and these factors may have an important influence in determining the nature of the processes operative within any given estuary. Following PRITCHARD (e.g. 1967), we can recognize four distinct estuarine types:

(*a*) *Drowned river valleys* (*or 'coastal plain estuaries'*) These have been formed by a rise in sea-level relative to the land, as a result of the release of

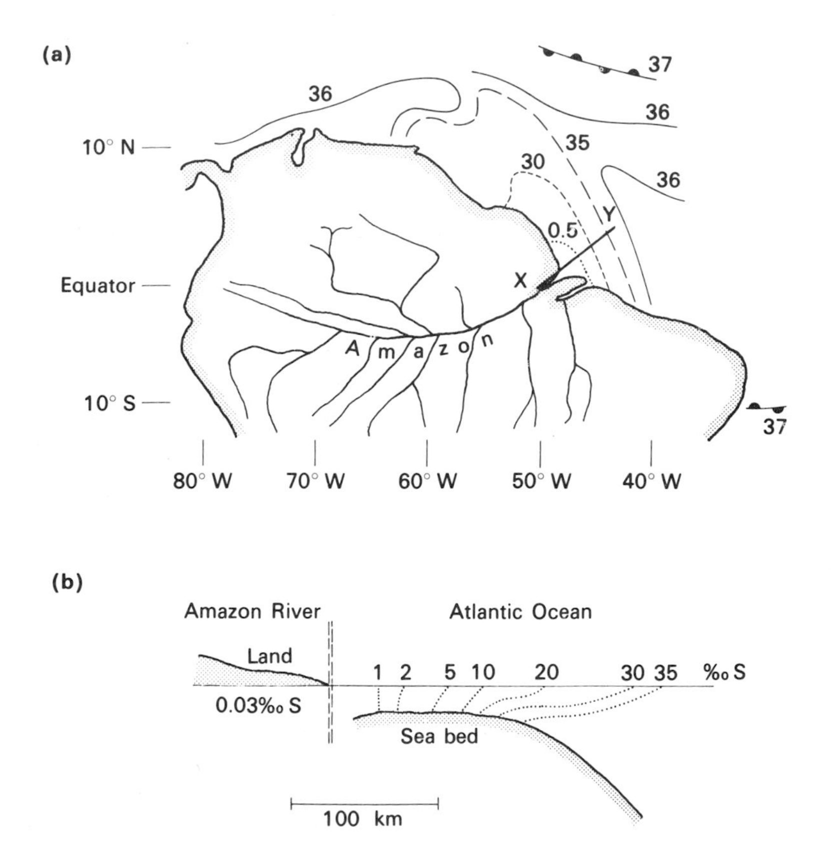

Fig. 1–1 (**a**) Surtace isohalines (‰ S) near the mouth of the Amazon. Note that the Amazon discharge is deflected north-westwards by the offshore Guiana Current (after various authors). (**b**) Vertical section approximately along the line X–Y in (**a**) (after GIBBS, R.J. (1970). *Journal of Marine Research*, **28**, 113–23).

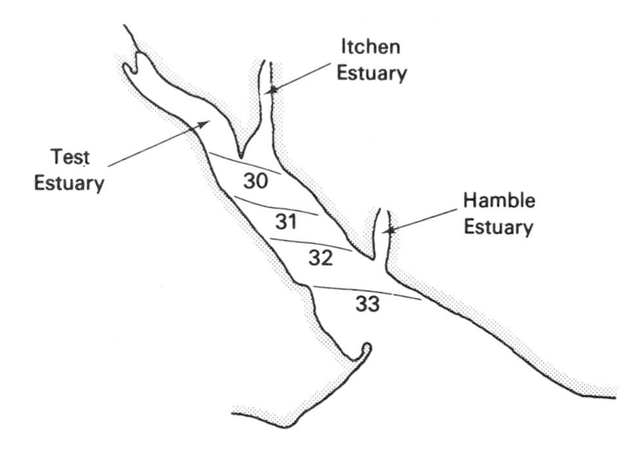

Fig. 1–2 Surface isohalines (‰ S) in Southampton Water (from a figure kindly prepared by K.R. Dyer).

ice-held water at the end of the last glaciation. They are the most common type of estuary and will be the ones receiving most of our attention later.

(*b*) *Tectonically-produced estuaries* The sea may invade the land not only as a result of rise in sea-level, but due to land subsidence. Members of this small class of estuaries originate from the invasion, by the sea, of fault-lines extending to the coast (e.g. San Francisco Bay). For our purposes, such estuaries approximate to those in drowned river valleys.

(*c*) *Fjords* Fjords are glacially-overdeepened valleys into which sea water now penetrates. Their special interest lies in the frequent presence of a sill at their mouth which may rise to within 200 m of the surface although the fjord itself can be over 1250 m deep. This sill may greatly restrict the interchange of water between the fjord and the adjacent sea, so that the volume of water lying beneath its level is effectively isolated and can occasionally be stagnant.

(*d*) *Bar-built estuaries* Some 13% of the world's coastline – although only 5% of Europe's – is dominated by offshore or along-shore barriers of sand or, more rarely, of shingle. These barriers may take the form of spits as a result of long-shore drift or of offshore bars or islands which may be moved towards the shore by, for example, storms. Such bars or spits can partially block and divert a discharging river system so that its estuary becomes located in a former area of sea enclosed between the sand/shingle barrier and the land; only if the freshwater discharge or tidal fluctuations are relatively large may complete enclosure by the barrage be prevented. These bar-built estuaries, for example those of the Ore/Alde in England, the Vellar in India, and the Indian River in Florida, merge insensibly with coastal lagoons (p. 67), and the difference between the various categories of coastal environment is here at its most arbitrary. A system is generally considered an estuary if its mouth is relatively wide and if the through-put or exchange of water is large in relation to its volume; conversely, if the mouth and the through-put or exchange of water are small, then the system is more realistically regarded as a lagoon. Several large bays receiving freshwater input have also been semi-enclosed by chains of barrier islands, for example the Waddenzee in north-west Europe and Moreton Bay in Queensland, and accordingly these regions also approximate estuarine or lagoonal habitats.

1.2 Characteristics of the estuarine environment

We can now consider the conditions usually found to be associated with an estuary, bearing in mind that such is their diversity that there will always be some estuaries that will not display all the various features to be described.

1.2.1 Salinity

Because one end of an estuary grades into fresh water (which arbitrarily may be assigned a salinity of < 0.5‰) and the other is open to the sea (with a salinity of ≏ 35 to 37‰), it follows that a salinity gradient of some form will exist along the estuary. This gradient is rarely as simple as that portrayed in Fig. 1–3, however, since factors such as the difference in density between fresh and salt waters, the topography of the estuary, the relative volumes of the participating

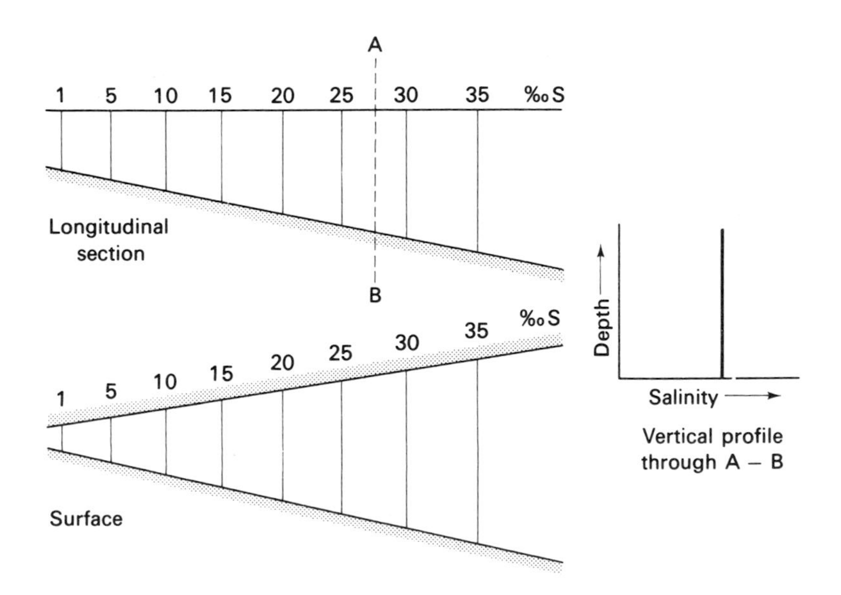

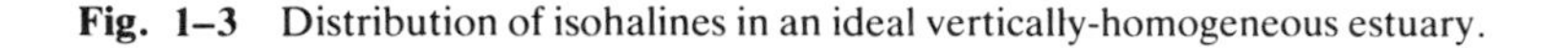
Fig. 1–3 Distribution of isohalines in an ideal vertically-homogeneous estuary.

waters, the rotation of the earth, etc., exert profound effects on the mixing processes. It is also important to realize at this juncture that the salinity of the water need not be intermediate between that of the adjacent sea and that of the inflowing fresh water. Where the volume of water lost by evaporation from the surface of the water mass exceeds that discharged into it by the rivers, the estuarine water can become hypersaline, i.e. its salinity may exceed that of the sea. Such a condition is not infrequent in low-latitude regions subject to seasonal droughts.

We can pass over the bar-built estuary and the fjord with the comment that, in so far as they differ from drowned river valleys, the former is often characterized by wind-induced mixing and the latter by restricted circulation, and concentrate on the mixing processes in the more common drowned river valleys. These processes will also operate in the surface waters of fjords. The dominant salinity regimes in such systems vary from a state of little mixing between the fresh and salt water masses, as shown by 'salt-wedge estuaries', through 'partially-mixed estuaries', to 'vertically-homogeneous estuaries' in which mixing is complete (as shown in Fig. 1–3). Although once again there is a continuum, we can conveniently examine the stratification and gradients of the two extremes and of the intermediate partially-mixed estuary as representatives of an estuarine series. The position of any given estuary in the continuum, i.e. the amount of mixing of the two participating water masses, in fact depends upon such factors as the river flow, the tidal velocity, and the width and depth of the estuary.

In a salt-wedge estuary, for example that of the Mississippi, salt water extends from the sea up the estuary in the form of a wedge below the less dense outflowing fresh water. Here the discharging river flow is sufficiently large to dominate the circulation system. If friction did not take place between the different water masses and between the sea water and the substratum, the interface between the two bodies of water would be horizontal and it would extend up the estuary to the point at which the bed was at sea-level. Frictional forces do exist, of course, and consequently the interface slopes downwards somewhat on progression towards the freshwater source (Fig. 1–4). As little mixing occurs, there is a distinct discontinuity between the salinity of the two water masses in vertical section (i.e. a marked halocline exists).

More commonly, mixing of the two water masses is effected by the breaking of internal waves at the interface between the fresh and the salt water, and by current-induced turbulent mixing, leading to the partially-mixed condition. Sea water is thereby introduced into the outflowing freshwater mass, increasing its salinity, and an upstream movement of sea water occurs along the bottom of the estuary to replace that lost. Compared with a salt-wedge estuary, the halocline is less sharp in vertical section (Fig. 1–5).

If tidal currents are strong enough to dominate the circulation system, the mixing processes become so complete that an estuary similar to that depicted in Fig. 1–3 can result. The Thames and the Severn Estuaries approximate to this condition.

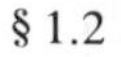

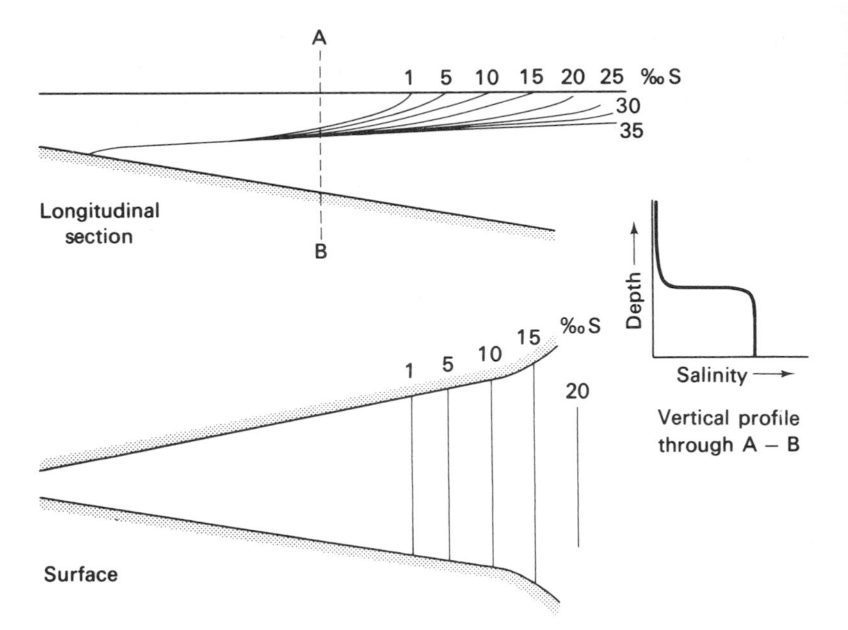

Fig. 1–4 Distribution of isohalines in a salt-wedge estuary.

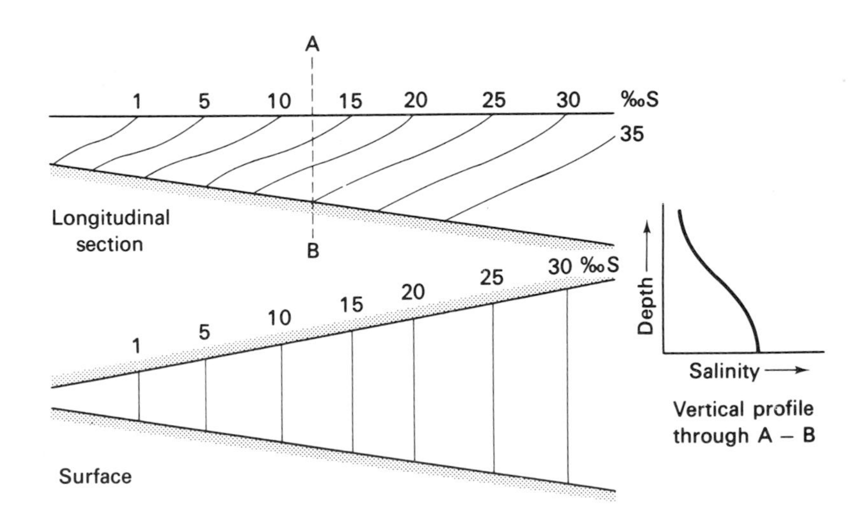

Fig. 1–5 Distribution of isohalines in a partially-mixed estuary.

On to these three types of estuarine mixing process may be superimposed Corioli's forces. The rotation of the earth has the effect of deflecting the outflowing fresh water to the right of an observer looking seaward and the inflowing sea water to the left (in the northern hemisphere). The salinity of the water at corresponding points on opposite sides of an estuary may therefore be very different. This phenomenon is shown on a large scale by Chesapeake Bay.

We may also superimpose two other factors: the effect of rise and fall of the tide in the adjacent sea and the effect of wet and dry seasons of the year. On an incoming tide, when sea water moves into an estuary, the zone of mixing will move upstream, as it will also do if the volume of discharged fresh water is reduced during a dry season. When the tide is ebbing, or during the wet season, the zone of mixing will move seaward. Therefore, at different states of the tide during a single day and at different seasons of the year, any given point on an estuary will be subjected to differing ambient salinities. The extent of these fluctuations will be damped, however, within the substratum (see below).

1.2.2 Substrata

The mixing of fresh and salt waters has other effects besides the establishment of some form of horizontal and/or vertical salinity gradient. The first of these which we shall consider concerns the deposition of fine particulate matter in estuaries. The details of the sedimentation process are complex (see, e.g. DYER, 1973) and depend upon such factors as water circulation pattern, topography, and the locally available sediments. Although these cannot be treated in detail here, we can investigate the basic pattern of sediment deposition in general terms.

By the time most rivers reach the sea, they have already deposited the coarser sediments in their upper reaches. Silt particles (which lie within the range 4 to 63 μm diameter) are transported in suspension in the lower reaches of most rivers, however, and will be discharged into the adjacent estuaries. On contact with a medium containing high concentrations of cations, these silt particles tend to flocculate; so that at the halocline (or in a vertically-homogeneous estuary, at the zone where 'pure' fresh water meets water of higher ionic strength) the silt particles borne by the fresh water clump together and sink more speedily. Of the three more important mineral fractions of clay in suspension, illite and kaolinite flocculate in salinities of less than 4‰, whereas the flocculation of montmorillonite continues over the whole estuarine salinity range. Flocculation and the fall velocities of the floccules are affected by temperature and by the amount of organic and inorganic matter in suspension, in addition to salinity.

Although the floccules tend to sink, they may be carried into outflowing fresh water by the circulation system, upon which they will de-flocculate and a flocculation/de-flocculation cycle can result. Some will reach and adhere to the substratum, however, although many will be resuspended by current action at ebb tide, and if the concentration of sinking floccules is very high (*c.* 10 g silt l^{-1} water) liquid mud may form which will flow as a layer close to the bottom. In most estuaries, net deposition exceeds erosion so that there is an overall accumulation of mud. It has been calculated that, as a general average, some 2 mm of mud accumulates in estuaries per year.

This description has assumed that most of the materials deposited in estuaries have been transported into the system by river flow. This is not always the case, however. In those estuaries in which sea water is the dominant water mass, for example the Mersey, the bulk of the deposited material is of marine origin. The tides bring large quantities of suspended mud into estuaries and under the prevailing comparatively sheltered conditions, the mud may settle out. This occurs particularly at high water slack tide over the littoral areas (any mud deposited at low water slack tide nearer mid-channel is liable to be swept away by current flow), leading to the development of extensive littoral mud-flats. This process is greatly accelerated by the presence of salt-marsh vegetation.

Thus, through the dual action of shelter and flocculation, estuaries tend to be regions in which fine sediments accumulate: most estuaries are muddy! But if the inflowing rivers drain hard igneous ground and if the offshore deposits are sand, stone or shell, the estuary may be predominantly sandy, for example those of north Devon and of Cardigan Bay. Once again, the danger of generalization on the subject of estuaries is apparent.

The deposited fine sediments display one particularly important feature: their interstitial water is less labile in its salinity than is the overlying water mass (Fig. 1–6). Many species can utilize this ameliorating effect of the substratum on the general fluctuating regime of estuarine salinities, thereby escaping from the worst rigours of the chemical environment.

1.2.3 Detritus

A system allowing the accumulation of fine inorganic particles will probably allow the settling out of particulate organic matter. Hence estuaries form efficient 'detritus traps'. The detritus may be swept in by the tides or it may originate from salt-marshes or mangrove-swamps along the shore line of the estuary (see pp. 19–21).

As we shall see later, this assumes great importance in the ecology of estuaries, particularly so when we consider another feature of the sedimentation process. We have noted above that estuarine waters carry high concentrations of silt in suspension and the water is therefore turbid. This may restrict light penetration to such an extent that algal production is reduced severely. Hence the abundance of detritus may be of greater significance as an alternative food source than it is in other intertidal or shallow-water habitat-types.

1.2.4 Wave action

Most estuaries take the form of a triangle of which the base is small and is open to the sea and of which the other two sides are long and are formed by

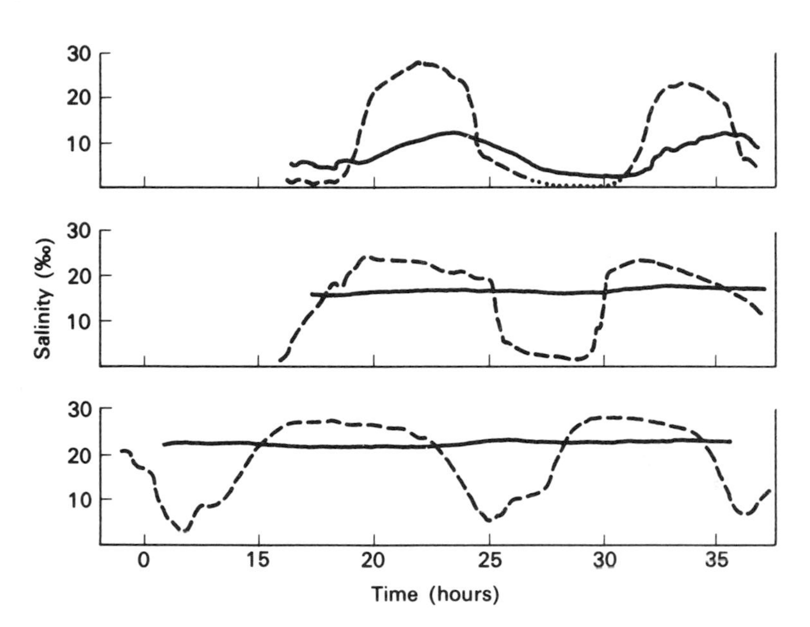

Fig. 1–6 Variation in salinity of the overlying water (– – –) compared with that of the interstitial water (———) at three stations in the Pocasset Estuary. (After MANGELSDORF, P.C. Jr, in LAUFF (1967). *Estuaries*. A.A.A.S. Publication No. 83, pp. 71–9; Copyright 1967 by the American Association for the Advancement of Science.)

land. Consequently the fetch of the wind is very short from most directions and therefore the height of waves is generally small. The only direction from which the fetch is usually of sufficient magnitude to generate large waves is along a line passing down the length of the estuary and out through its mouth. Diffraction of the waves occurs, however, on passage through the mouth and, in addition, the mouth of many estuaries is partially closed by a sand or shingle spit, which further increases the diffraction. The energy of these waves is therefore largely dissipated before the main body of the estuary is reached. The narrowness of the mouth of many estuaries also results in a dampening of tidal action and a reduction in the speed of tidal oscillation, although water velocities may be high in the mouth itself.

1.2.5 Other characteristics

The nature of the salinity gradients and of the sediments, the abundance of detritus and the relative freedom from wave action are probably the four most important features of the estuarine environment from the point of view of most organisms. We will now briefly mention a number of other important variables.

(*a*) A second effect of the mixing of two different water masses in estuaries can be seen in the chemical composition of estuarine water. Sea water and river

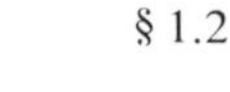

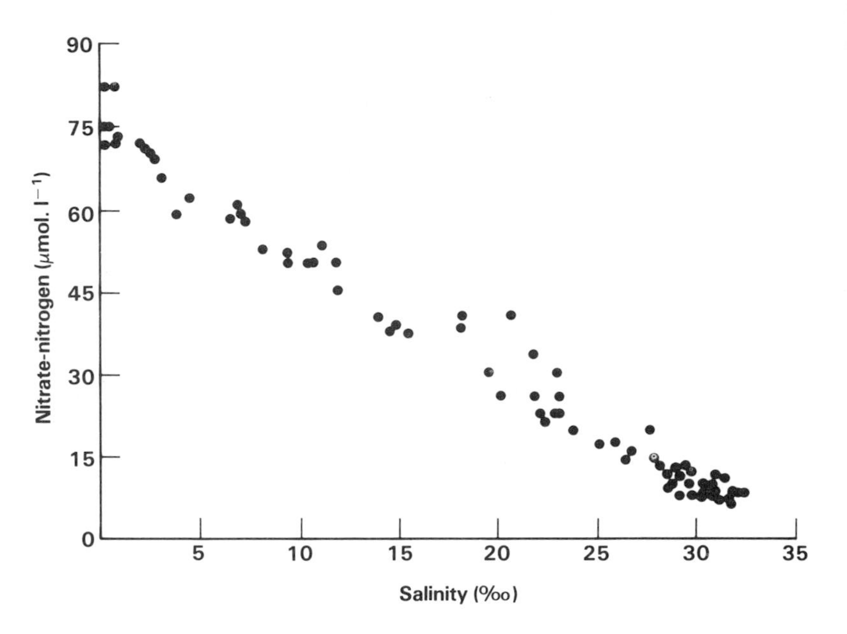

Fig. 1–7 The negative correlation between nitrate concentration and salinity in estuaries. (Data from the Wyre Estuary supplied by the North West Water Authority through the kindness of Dr P.C. Head.)

water do not possess inorganic ions in the same relative proportions. Sea water, for example, possesses less bicarbonate and silicate (expressed as SiO_2) and more chloride (as percentages of the total dissolved inorganic matter) than river water. Phosphate and nitrate, in particular, are negatively correlated with salinity in estuaries (Fig. 1–7). The ionic form in which various chemical elements are present also differs markedly between fresh and salt waters and this may have a bearing on their availability to organisms. In estuaries, therefore, both the form in which chemical elements occur and the total amount present per unit volume will depend upon the water budget of the particular estuary.

(*b*) The shores of estuaries are subjected to wide fluctuations in temperature when exposed at low tide. During daytime low tides in summer, the temperature of the mud-flat surface is liable to be highest, whereas it will be lowest during night-time low tides in winter. In particularly cold winters (e.g. that of 1962/63), the surface waters of several estuaries freeze. Such temperature fluctuations, however, are not specifically characteristic of estuaries: they are a general feature of intertidal regions.

(*c*) Microbial decomposition of the large quantities of detritus in estuarine sediments exerts a heavy oxygen demand on the interstitial water. Moreover, owing to the reduced lability of this interstitial water, any oxygen removed cannot be quickly replaced. Even within the uppermost 5 to 10 mm of the sediment little oxygen may be available and below this depth the substratum is usually anoxic. In such situations, the burrows of actively irrigating animals are clearly visible as brown tubes in a black anoxic matrix. In some estuaries with slow flushing rates, the overlying water mass also exhibits low oxygen tensions as a result of the amount of organic matter in suspension. The most frequent cause of this state, however, is pollution (see pp. 51–7).

1.3 The life-span of an estuary

In terms of the earth's geological history, individual estuaries are transient phenomena. Although some may have a history to be counted in millions of years, many estuaries extant today are the products of the rise in sea-level consequent on the retreat of the last glaciation, and as such they are only some 8000 years old (in anything approaching their present form). Extending a similar length of time into the future, and assuming a constant sea-level, some existing estuaries will be extinct.

Basically, this will result from sediment accumulation and stabilization, although over the last few decades man has assumed an active role in this process, with land-reclamation and freshwater reservoir projects. As sediment accumulates, it is invaded, stabilized and its deposition rate is accelerated by salt-marsh formation, confining the area of free water within narrowing limits. Simultaneously, long-shore drift may, as we have seen, form a bar across the mouth of small estuaries, which can ultimately exclude the inflow of sea water. The resulting lagoon will rapidly develop into a swamp or into marshland, and finally into the climax vegetation typical of its geographical and climatic region. Examples of this process can be seen along the south-western coast of England, the southern coast of the U.S.A., in Western Australia and in many other regions.

Conversely, erosion may widen the mouth of an estuary, so that it gradually loses its separate identity to become a shallow coastal bay. The discharged fresh water will not then be confined within such narrow limits and, unless the freshwater volume is large, marine currents will progressively dominate the area.

2 Nature of the Fauna and Flora

2.1 Generalities

If an estuarine fauna is compared with those in the adjacent sea and in the inflowing river systems, it will be seen that the estuarine fauna is generally poor in numbers of species, although it may be rich in numbers of individuals. This is partly due to the inability of many freshwater species to inhabit more saline media and of many marine species to withstand dilute media. Very few freshwater species survive in salinities in excess of 5‰ and few marine species are to be found in salinities of less than 18‰. Of course, benthic marine animals can penetrate salt-wedge estuaries to a considerable extent. A similar phenomenon is to some degree manifested by the flora. The number of emergent phanerogams decreases along a fresh water → brackish → marine series, whilst the algal flora is generally richest in the sea.

Besides the reduction in numbers of marine species observed within an estuary (Table 2–1), a number of morphological and physiological changes occur in those few species which do penetrate this environment. Many marine

Table 2–1 Progressive reduction in numbers of species with distance into the Baltic* (from data of Remane, in REMANE and SCHLIEPER, 1971, by courtesy of E. Schweizerbart'sche Verlagbuchhandlung).

	North Sea (30–35‰S)†	*Kiel Bay* (13–20‰S)†	*South and central Baltic* (6–8‰S)†	*North Baltic* (4–6‰S)†
Porifera	64	18	1	0
Hydroida	82	34	7	3
Ectoprocta	*c.*90	35	5	2
Polychaeta	*c.*250	*c.*100	22	4
Amphipoda	147	55	20	9
Decapoda	*c.*50	12	6	2
Prosobranchia	114	26	13	1
Lamellibranchia	92	32	11	4
Echinodermata	39	10	2	0
Ascidiacea	24	7	1	0

*The Baltic Sea can be considered a large stable estuary
†Surface salinities

species of both animals and plants attain a smaller size in estuaries than in the sea (Fig. 2–1) and show a reduction in the number of the units comprising meristically varying characters (e.g. vertebrae). This dwarfism, however, is not displayed by predominantly freshwater species. The reproductive rate of marine species is often lowered in brackish conditions, for example the reproductive season may be delayed and/or its duration reduced and fewer eggs may be produced. Freshwater organisms are also frequently semi-sterile in estuaries. Such changes occur in populations which may otherwise appear to be 'successful', as evidenced by their abundance and biomass.

The other factor contributing to the reduced number of species in estuaries is the paucity of the characteristic brackish-water fauna. This can never effectively replace the missing freshwater and marine components (Fig. 2–2). We shall consider the nature of the brackish-water components below.

Before so doing, however, it is worthwhile considering why there are so few species in this brackish-water fauna, in spite of the abundant food supply which permits those few species which do exist to achieve high densities. The standard answer to this question is to point to the rigorous nature of the physical environment. On this basis, only a limited number of species are capable of evolving the physiological specializations to enable them to withstand brackish water of fluctuating salinity. Several marine groups, for example the echinoderms, find this difficult to achieve. Other workers point to the short time-span of estuaries and suggest that animals do not have sufficient time to adapt and speciate before their habitat vanishes, i.e. the estuarine fauna is essentially opportunistic.

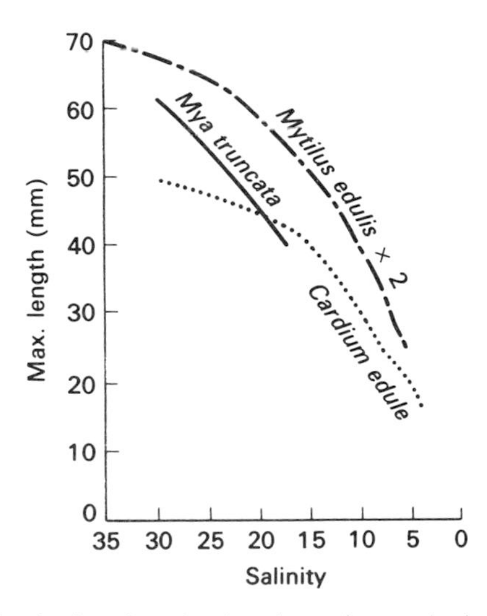

Fig. 2–1 Reduction in size of predominantly marine species in low salinities. (After REMANE, A. (1934). *Verhandlungen der Deutschen Zoologischen Gesellschaft*, **36**, 34–74.)

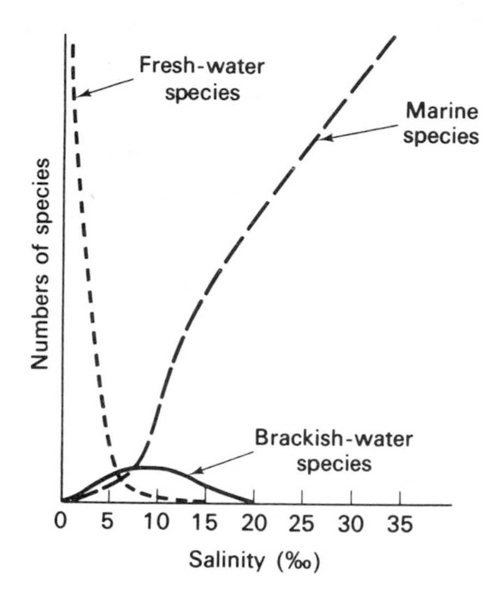

Fig. 2–2 Graph of numbers of (*a*) freshwater, (*b*) marine, and (*c*) brackish-water species in different salinities. (After REMANE, A. (1934). *Verhandlungen der Deutschen Zoologischen Gesellschaft*, **36**, 34–74.)

Probably of even greater significance is the general observation that when an environment or area contains very few habitat types (and when those that it does contain show little spatial heterogeneity, i.e. are uniform and monotonous), then will its fauna be poor in number of species. The main habitat available to animals in estuaries is mud-flat and although the particle size, degree of shelter, etc., of mud-flats vary, they are variable within rather narrow limits. Hence the number of niches will, of necessity, be small. This is clearly shown by investigation of shores bearing stones and rocks on the surface of the mud. These stones are frequently covered by a number of sedentary and sessile species otherwise absent through lack of available substrata. The environmental uniformity of estuarine shores will then be sufficient in itself to set a comparatively low limit to the maximum numbers of species which can occur. Other factors contributing to the paucity of the fauna are considered in the next chapter (p. 35).

2.2 Benthos and benthic feeders

2.2.1 Macrofauna

The macrofauna comprises two ecologically distinct groupings: those (mainly invertebrate) species permanently resident within an estuary, and those (mainly vertebrate) species entering estuaries at high or low tide, principally to feed (e.g. fish and birds respectively).

(i) RESIDENT SPECIES The resident species of temperate regions are drawn mainly from within the Polychaeta, the Mollusca (especially lamellibranchs) and the Crustacea (especially isopods and amphipods). In tropical and subtropical areas, the dominant animals are gastropod molluscs and crabs of the families Ocypodidae and Grapsidae. Crabs, however, are infrequent in temperate estuaries; for example in north European estuaries the only native crab to occur commonly is *Carcinus maenas*, whereas in the estuary of the Brisbane River in Queensland, some 25 brachyurans are present.

It is the resident invertebrate macrofauna of estuaries which has received most study and hence we will consider the nature of the brackish-water component of the fauna using examples from this group. The fauna of temperate estuaries (excluding the predominantly marine and freshwater species) conforms quite closely to the '*Macoma balthica* community' of Petersen (see, e.g., THORSON, 1957). Estuaries are therefore characterized by species such as *Macoma balthica* itself, *Cardium glaucum* (and frequently *C. edule*), *Scrobicularia plana* and *Mya arenaria* (lamellibranchs), *Hydrobia ulvae* (prosobranch), *Nereis diversicolor, N. virens, Nephtys hombergi, Pygospio elegans, Manayunkia aestuarina* and *Arenicola marina* (polychaetes), and by *Corophium volutator, Gammarus* spp. (e.g. *zadduchi, duebeni* and *chevreuxi*), *Idotea chelipes, Sphaeroma* spp., and *Neomysis vulgaris* (crustaceans). Some of these species are illustrated in Figs 2–3 and 2–4. Towards the freshwater end of the estuarine salinity gradient, tubificid oligochaetes (e.g. *Tubifex* spp. and *Limnodrilus hoffmeisteri*) often dominate the fauna, and one species, *Tubificoides benedeni*, occurs in abundance in up to fully saline conditions. Oligochaetes frequently are especially important elements in the fauna of polluted regions (MCLUSKY *et al.*, 1981).

We must now ask to what extent these are specifically brackish-water or estuarine forms, i.e. do they require brackish water *per se*, are they adapted to the whole gamut of estuarine conditions, or do they inhabit estuaries for other reasons? Excluding the epifaunal crustaceans from present consideration (see p. 44), evidence is rapidly accruing to suggest that none of these species is in fact specifically estuarine, although together they form a characteristic estuarine fauna (see, e.g. BOYDEN and LITTLE, 1973). That is to say that the brackish-water component of Remane does not apply in practice to the estuarine situation, although of course it may still be relevant to other brackish areas, for example the brackish seas of Europe and Asia.

Macoma, Hydrobia, Corophium, Nereis diversicolor and *Cardium glaucum* appear to be fundamentally sheltered-habitat species; for example *Corophium volutator* occurs sublittorally in marine situations where mud is present and the water is quiet (e.g. in parts of the eastern Solent), and can survive salinities of 2 to 50‰ (MCLUSKY, 1971) provided these conditions are satisfied. In *Cardium glaucum*, at least, the limiting effects of wave action are felt by the newly settled spat (BOYDEN and RUSSELL, 1972). Yet other species, for example *Scrobicularia, Nephtys, Arenicola, Manayunkia* and *Mercierella enigmatica* (see below), are substratum-specific. *Manayunkia*, for example, hitherto

Fig. 2–3 Estuarine crustaceans and polychaetes: top left, *Idotea chelipes*; bottom left, *Corophium volutator*; top right, *Nereis diversicolor*; bottom right, *Arenicola marina*. Scale rod = 1 cm. (Specimens from Norfolk photographed by N.C. Maskell.)

Fig. 2–4 Estuarine molluscs: top left, *Hydrobia ulvae*; top right, *Cardium glaucum*; bottom left, *Macoma balthica*; bottom right, *Scrobicularia plana*. Scale rod = 1 cm. (Specimens from Norfolk photographed by N.C. Maskell.)

regarded as restricted to estuaries, is now known to occur in soft detritus in fully marine regions.

Of course estuaries are areas notable for their high degree of shelter from wave action and for the deposition of fine sediment. Along many coastlines they are the only areas where such conditions prevail. But mud and shelter do occur elsewhere (e.g. to the lee of fringing islands) and the faunas of these areas are indeed those normally associated with estuaries. Estuaries are, however, specifically areas where fresh and salt waters mix and it would appear that many, if not all, 'estuarine species' do not require such a salinity regime; they merely tolerate it to differing degrees in order to inhabit the food-rich mud and to gain shelter from wave action.

As we shall see later, estuaries form ideal natural harbours and hence they are used by long-distance commercial shipping. This agency has introduced another component into the fauna: the immigrant species. *Elminius modestus*, *Eriocheir sinensis*, *Styela clava*, *Mercierella*, etc., are today abundant in many north-temperate estuaries and, where the water is warmed by the discharges from coastal power stations, *Mercenaria mercenaria*, *Brachynotus sexdentatus*, *Bugula neritina* and *Balanus amphitrite* may occur. *Crepidula fornicata* and *Petricola pholadiformis* have also established themselves in temperate European estuaries, after their original introduction with stocks of American oysters. In Southampton Water, the biomass of accidentally introduced species probably exceeds that of the native species.

(ii) MIGRATORY AND NEKTONIC SPECIES The majority of the migratory fauna uses estuaries as feeding grounds and hence the biology of these species will be considered in more detail in the next chapter. The organisms involved are principally fish and birds.

Very many families of fish have representatives which feed in estuaries during some stage of their life cycle, for example the Engraulidae, Clupeidae, Mugilidae, Sciaenidae, Gobiidae and Pleuronectidae, and extensive fisheries for some species, for example menhaden (*Brevoortia tyrannus*), have developed. There is even a possibility that estuarine fish farms (for crustaceans and molluscs as well as fish) may help to meet food shortage in many parts of the world. Other fish, for example bay anchovy (*Anchoa mitchilli*), enter estuaries to spawn, whilst anadromous and catadromous fish pass through estuaries during their spawning migrations (in opposite directions).

The number of birds feeding in estuaries is very large. Waders and gulls (Charadriiformes) and wildfowl (Anatidae) are the most important groups, but several of the Pelecani and Ciconiiformes are represented in warmer regions. Several of the species also use estuarine salt-marshes and mangrove-swamps as roosts.

2.2.2 *Microorganisms and meiofauna*

Only a small number of estuaries have been investigated with respect to their microfauna. Such studies as have been undertaken indicate a fauna rich in species and individuals of ciliates, nematodes, ostracods and harpacticoids. Turbellarians (especially rhabdocoels), a few very specialized coelenterates (e.g. *Protohydra* and *Nematostella*), rotifers, gastrotrichs, archiannelids and halacarid mites also occur. The richness of this fauna therefore contrasts markedly with the paucity of the macrofauna. Such an increase in species diversity is probably attributable to several factors. Microscopical examination of substratum samples from estuarine mud-flats reveals a much greater spatial heterogeneity, for example, than is apparent from the comparatively gross viewpoint considered earlier (p. 14), and decreased salinity fluctuations are typical of the habitat of the microfauna (Fig. 1–6). Several microfaunal species appear to be restricted to estuaries, but it is not yet known to what extent a truly endemic fauna exists.

Flagellates, diatoms and blue-green algae may all be abundant in and on estuarine muds. Several of the algae are motile (e.g. euglenoids and several diatoms) and these migrate through the top few millimetres of the substratum to the surface and back again in a diurnal rhythm. It appears most likely that they use a light stimulus to initiate this migration. Other species form mats on the mud surface, several species (including both diatoms and blue-green algae) being present in the same mat. The filaments of *Ulothrix*, *Lyngbya* and especially *Vaucheria* exert a profound stabilizing influence on the substratum surface, as does the mucus released by diatoms. Benthic diatoms are frequently far more numerous in an estuary than their planktonic relatives, although the distinction between 'planktonic' and 'benthic' here is somewhat vague as many benthic diatoms will be put into suspension by the incoming tide.

The water and mud of estuaries are extremely rich in bacteria. ZOBELL and FELTHAM (1942) found that the water over a mud-flat contained 200 times more bacteria, on average, than did the inflowing sea water (an average of 300 000 bacteria ml^{-1} as opposed to 1500 ml^{-1}). The surface layers of the mud, i.e. the top 5 cm, contained up to 1500 times more bacteria than did the bacteria-rich overlying water, with values of between 170 and 460 million bacteria per gram of substratum. These bacteria live in the water in the interstitial spaces between the particles and, to a lesser extent, on the surfaces of the particles themselves (most particles are usually devoid of bacteria and less than 1% of the surface area of colonized particles is occupied by attached individuals). Their metabolism is predominantly respiratory and they therefore exert a heavy oxygen demand on the interstitial water. In the absence of free dissolved oxygen, some bacteria can use the abundant sulphate dissolved in sea water as an oxygen source to permit a respiratory metabolism to be maintained under anoxic conditions. Bacterial reduction of sulphate results in the release of sulphide and accounts for the sulphurous nature of estuarine muds beneath the thin, brown, oxidized surface layer. Decaying mats of salt-marsh vegetation are also rich in bacteria: 20 million ml^{-1} have been recorded from *Spartina* debris, together with hundreds of thousands of diatoms and heterotrophic and photosynthetic flagellates in the same volume. As with the sediment particles, so in the *Spartina* mats most bacteria ($>$ 70%) are located in the interstitial water, more than half of the individual pieces of debris lack attached bacteria, and less than 20% of the surface area of colonized particles are occupied (WIEBE and POMEROY, 1972).

2.2.3 *Macroflora*

The macrophytes of estuarine situations have been much studied (see, e.g., WAISEL, 1972), but space here must restrict us to a very brief consideration of their general biology.

Below mean sea-level in both temperate and tropical estuaries occur a number of genera of 'sea-grasses', for example *Zostera, Cymodocea, Thalassia* and *Ruppia*. These are not only important primary producers in their own right, but bear a rich epiphytic flora of algae ('aufwuchs').

A characteristic series of salt-marsh plant communities occurs on the upper half of the shore in temperate regions. These phanerogams colonize the bare mud surface, for example such genera as *Salicornia, Arthrocnemum* and *Spartina*, from water-borne seeds or vegetative fragments, and in so doing accelerate the rate of sediment deposition by reducing the rate of flow of the tidal water. At Scolt Head Island, accretion rates during salt-marsh development have been found to equal the deposition of some 20 cm of mud in as many years, and values of up to 20 cm per year have been recorded in France. As the height of the mud-flat increases, other species requiring a longer period of aerial exposure for successful germination can become established and a series of zones of increasing height above the original mud-flat level may form. In this way, salt-marsh may reclaim mud-flat and convert it into pasture or scrub. The precise species involved in the succession from bare mud (with only an algal cover if any) to *Juncus, Scirpus*, etc., and ultimately (if the natural succession is allowed to proceed unchecked) to carr and woodland vary with a whole complex of factors, including the nature of the substratum, the existing species, the geographical area, etc., and numerous local zonations and successions have been described.

The fact that many of the salt-marsh species are long-lived perennials, i.e. with a life-span of up to 40 or 50 years, may render the succession extremely complex. The life of such plants may span much of the 'life' of the marsh and relations between the root-systems of adult plants and the surrounding soil may inhibit the successful germination of other seedlings. Once established, however, most salt-marsh plants spread vegetatively, seed being rarely set.

Similarly to the *Zostera* meadows (and to mangrove-swamps), the emergent salt-marsh phanerogams bear a large associated algal flora, whose ecological importance may rival that of the larger and more obvious species in some circumstances.

Salt-marshes develop extensive drainage-creek and salt-pan systems, in which many of the animal species characteristic of the adjacent mud-flats may live. In particular, the pans often possess a distinctive fauna including relatively large numbers of insect species (probably as a result of the even greater reduction in wave action). The fauna of the marsh surface contains a large terrestrial component.

In tropical and subtropical regions, mangroves or *Nypa* swamp ecologically replaces the temperate salt-marsh. Mangroves are shrubs or trees, of up to 30 m in height, belonging to several unrelated families but showing convergent evolution to life in unconsolidated and often anaerobic muds (cf. salt-marsh vegetation). Mangrove species show a number of physiological and morphological adaptations with respect to their gaseous exchange, nutritive, dispersal and support systems, which have been discussed at some length by MACNAE (1966). The parallels between mangrove and salt-marsh species are many: mangroves also show distinct zonations, both along the river–sea gradient and up the shore, and they are also efficient converters of intertidal mud into land (at a rate of over 100 m $year^{-1}$ in parts of Sumatra – cf. rates of up to 50 m $year^{-1}$ in European salt-marshes); both show a succession of different plant communities through time, with the pioneer colonists – frequently *Avicennia* in mangrove-swamps – progressively being replaced by other species associations; both are highly productive stands of vegetation (fixing some 2.7–5.5gC m^{-2} day^{-1} in comparison to a maximum of less than 0.5gC m^{-2} day^{-1} for the most productive of estuarine phytoplanktonic communities); both may export much of this production (some 0.9–1.3gC m^{-2} day^{-1} from salt-marshes and 0.5–2.4gC m^{-2} day^{-1} from mangrove-swamps) to adjacent aquatic environments; the plants comprising both show parallel adaptations including dispersal mechanisms utilizing sea-water transport; both possess a fauna containing a large terrestrial component and provide roosts for the estuarine avifauna; and both are drained by extensive and dendritic drainage-creek systems which, from the air, show an identical physiographic pattern. The parallel nature of salt-marshes and mangrove-swamps is even such that in the drier regions of the tropical mangrove belt, where their development is stunted, a form of salt-marsh – including *Arthrocnemum* spp. – replaces parts of the mangrove forest, especially at high tidal levels.

Salt-marsh and mangrove-swamp do, however, show a number of important differences that are mainly consequent on the arboraceous nature of mangroves and the closely-knit, turf- or meadow-like character of salt-marsh. Mangroves, for example, provide hard surfaces to which sessile estuarine animals may attach and they leave extensive areas of bare sediment between their trunks in which mud-flat animals may burrow (these habitats are rare or absent on salt-marsh); mangrove-swamp extends further down the intertidal zone than do many salt-marshes, mangroves almost reaching mean tide level but salt-marsh scarcely extending below the high-water neap-tide mark in North-west Europe (American *Spartina* marshes, however, may extend over the same vertical range as mangrove-swamps); and the zonal distribution of mangrove species imparts a zonation to the associated aquatic fauna, which in mangrove-swamp, as on the mud-flat at lower tidal levels, is dominated by decapod crustaceans, gastropod molluscs and periophthalmid fish (as we have seen, an extensive aquatic fauna is absent from the salt-marsh surface). Mangroves also provide a habitat for a rather more exotic estuarine fauna – at least to a temperate-zone biologist – including fire-flies, amphibians (the crab-eating frog), reptiles (various crocodilians) and mammals (leaf and proboscis monkeys). Through their conversion of intertidal regions into land, mangrove-swamps may have provided one well-used route for the colonization of land by marine organisms (LITTLE, 1983).

2.3 Plankton

The plankton of estuaries is usually sparse in numbers of species, although frequently rich in numbers of individuals: a phenomenon we have also noted in the estuarine benthos. In some estuaries (mainly in the tropics), however, at least the diatom flora is characterized by the presence of many species, each with a comparatively small population.

As in lakes and in the sea, an annual cycle of phytoplankton abundance is present in estuaries. During the last quarter of the year, phytoplankton numbers are low, probably as a result of reduced light intensities and increased turbidity. A bloom occurs in late winter or spring in which the numbers of diatoms increase rapidly, the increased light intensity triggering the bloom. During the summer months, grazing pressure exerted by copepods, and possibly the depletion of essential nutrients by the bloom, causes decreased phytoplankton densities, although a second bloom occurs in some estuaries in the autumn. Diatoms (especially *Skeletonema, Asterionella, Chaetoceros, Nitzschia, Thalassionema* and *Melosira*) frequently dominate the phytoplankton, but small dinoflagellates (including *Gymnodinium* and *Gonyaulax*), often show irregular and localized blooms through the year and in many estuaries this group is dominant. The degree of turbidity and the water circulation pattern play marked roles in determining the nature of the phytoplankton composition and abundance, and consequently estuaries differ considerably with respect to the amount of phytoplankton production.

Maintenance of an endemic zooplankton population obviously depends on the flushing time of an estuary and on its circulation system. The zooplankton of estuaries is therefore also variable, both in species composition and in biomass, but numbers are generally high in summer following the late winter/spring phytoplankton bloom. Species of *Acartia, Eurytemora* and *Pseudodiaptomus* are characteristic estuarine zooplankton, and it is interesting to note that *Acartia* has been shown to be an inefficient feeder in comparison with coastal marine copepods. This suggests the utilization of a habitat through comparative freedom from competition. Coastal cladocerans, i.e. *Podon, Evadne* and *Penilia*, are frequently abundant in estuaries and several marine copepods can penetrate estuaries to some extent, *Temora, Pseudocalanus* and *Centropages* being perhaps the most common. Centropagid copepods are abundant in several estuaries in the southern hemisphere.

3 Food and Food Webs

3.1 'Detritus' and microorganisms

3.1.1 Sources of organic detritus

Basically, the food sources available to the primary estuarine consumers are threefold: phytoplankton, benthic algae and detritus. Detritus, however, is not really a single substance. It is a composite term embracing the fundamental decaying organic debris, the various microorganisms (fungi and, especially, bacteria) responsible for its decay, those microscopic organisms such as heterotrophic flagellates, ciliates and nematodes which consume the agents of decay, their meiofaunal predators, and miscellaneous other organisms, including diatoms and photosynthetic flagellates, which occur in the detrital interstitial spaces or which use the debris as a convenient substratum. An animal consumer ingesting a detrital flake or floccule may well take in all these different living and dead components, including some more properly belonging to the benthic algal category.

Phytoplankton is consumed by the estuarine zooplankton, often relatively inefficiently. In some highly productive estuarine regions, as much as 50% of the net phytoplanktonic production may remain ungrazed, and this then becomes available to the benthos in the form of dead and dying cells. The faecal pellets of the zooplankton also contain much unassimilated material. In both cases, the planktonic production includes a large portion contributing to the general estuarine pool of detritus. In addition, benthic suspension feeders such as mussels (p. 27) remove phytoplankton from suspension in the overlying water and deposit some of its contained organic matter on to the sediments in the form of faeces and 'pseudofaeces' (see p. 26).

The mats of algae on the sediment surface are grazed directly by some invertebrates, for example the small opisthobranch molluscs *Alderia* and *Limapontia*, and by one fish (the grey mullet), but these mats also contribute to the detrital pool, water movement tearing them and rolling detached fragments into pellets. These and the other benthic algae have average productivities in the order of 0.1–0.6gC m^{-2} day^{-1}. Further sources of algal food include the finer of the salt-marsh algae and the aufwuchs (p. 19) growing on eel-grasses and other submerged or semi-aquatic macrophytes.

Most of the organic matter available to estuarine invertebrates, however, is that included under the general category of detritus. Besides the sources of organic debris mentioned above, it originates from river-borne flotsam (leaves and branches of riverside trees, etc.), from material such as seaweeds ripped from rocky shores during storms and carried into estuaries by the tide, and from the fringing salt-marshes or mangrove-swamps. Different estuaries vary markedly in which source is the most important: those of the Rhine/Meuse Estuary in the Netherlands, for example, received half of their total input of carbon in the form of an import from the North Sea; estuaries in the south-eastern U.S.A. are mostly dependent on material originating from the fringing *Spartina* marshes; and some estuarine areas in New England are fueled mainly by *in situ* micro-algal productivity. This variation results from differences in predominant wind directions, in tidal ranges, and in the areal extent of fringing angiosperm communities.

Much interest, and considerable controversy, has been associated with the role of salt-marshes in estuarine productivity patterns. Some 'classic' marshes, including those in the south-eastern U.S.A. referred to above, export over 50% of their net productivity to the adjacent estuaries and to the sea (WEIGERT, 1979). One sample salt-marsh creek in Georgia was estimated to conduct 140 kg of organic matter from its 10–25 ha drainage basin out of the marsh each spring tidal cycle, and an equivalent 25 kg during neaps (ODUM and CRUZ, 1967). In this case, the marsh was exporting *Spartina* and macro-algal debris, but was retaining plankton brought on to it by the flooding tide. Not all marshes appear to produce such an export, however; others are more nearly in balance, and some may even prove to be net importers of fixed organic matter (NIXON, 1980). One New England marsh is known to be a net exporter of macroscopic material but a net importer of finely particulate debris.

3.1.2 *Colonization by microorganisms*

Whatever may be the original source, estuarine mud-flats and waters receive detrital materials mainly in the form of decaying leaves and stems of angiosperms and the fronds of macroscopic seaweeds; this material settles out on the mud surface and becomes incorporated into the interstitial spaces and adsorbed on to the particles. Not surprisingly, most of the organic matter in the sediments is therefore present on or near the surface (Fig. 3–1), although sedimentation may bury material before it has achieved complete remineral-

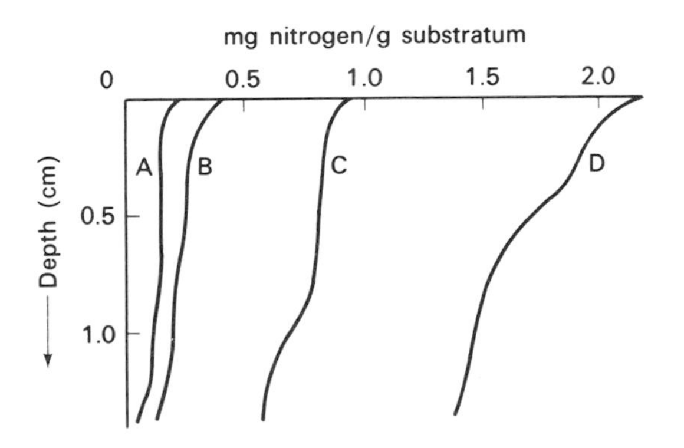

Fig. 3–1 Vertical distribution of organic matter in various substrata: **A**, sand; **B**, sandy mud; **C**, mud; **D**, silt. (After ONO, 1965.)

ization, so that estuarine sediments may be detritus-rich even at depths of several metres. At depth, obligately anaerobic bacteria may slowly continue the decomposition process. The quantity of detritus associated with the sediment varies with the slope of a mud-flat and, above all, with its degree of exposure to wave and current movement. In the most sheltered of regions, both fine silts and light organic particles can settle out of suspension; under regimes of greater water movement, more of the light materials, both organic and inorganic, will be retained within the water mass and only the heavier particles will sediment out. Hence there is usually a good inverse correlation between the organic content of estuarine sediments and their particle size in terms of modal particle diameter and proportion of fine particles in the sediment (Fig. 3–2).

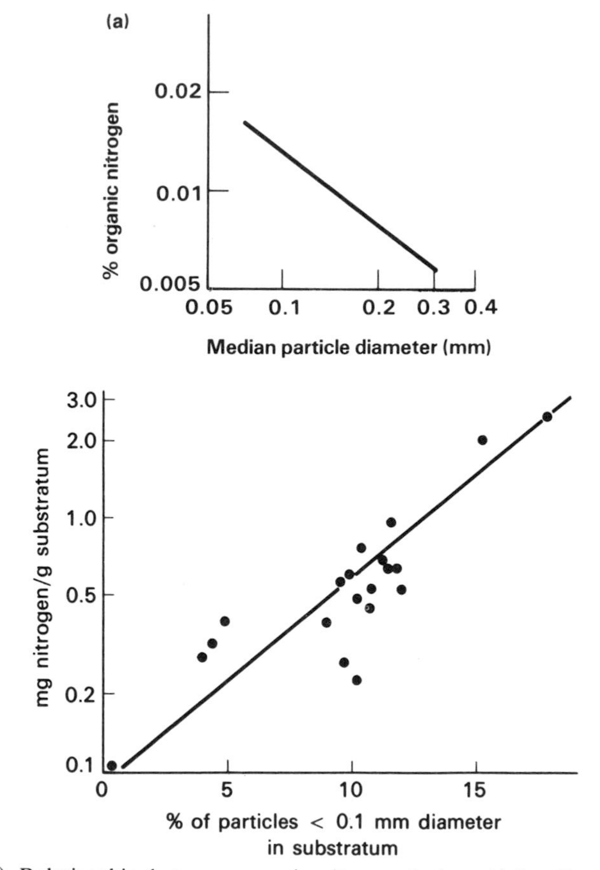

Fig. 3–2 (**a**) Relationship between organic nitrogen in intertidal sediments and median particle diameter (after LONGBOTTOM, M.R., unpublished). (**b**) Correlation between organic nitrogen content and quantity of fine particles in various substrata (after ONO, 1965).

If a microbial population had not already developed in association with the detrital particles whilst they were in suspension, such soon occurs on their deposition on to the mud, and animal faeces are similarly colonized. Bacteria appear first to colonize the interstitial waters, there presumably subsisting on dissolved organics leached from the detritus. Only later, when the soluble compounds have been exhausted, do they begin to occur on and below the surface of the debris and to utilize the more refractory insoluble compounds. As bacterial numbers begin to increase so do populations of bacteria-consuming flagellates and ciliates, in turn followed by meiofaunal nematodes, harpacticoids, etc. which feed on a variety of items lower down in the detrital food-chain. Algal numbers also increase.

Decaying plant material soon becomes leached of soluble nutrients (if, indeed, they were not translocated back into perennial tissues before the dead leaf or stem was shed) and hence leached detritus is nutrient-poor. Yet bacteria have a high demand for such nutrient elements as nitrogen and phosphorus. They may be able to obtain their required carbon from the detritus, but they must get their nutrients from elsewhere. This can only be from the interstitial water and, as a result of the low porosity and permeability of muds, this water mass may soon become depleted of nutrients, as the diffusing supply cannot keep pace with demand. Local nutrient shortage – not, as was once thought, lack of space on the particles for colonization (p. 18) – would appear to limit bacterial productivity and therefore the rate of decomposition of the detritus (FENCHEL and HARRISON, 1976). Many bacterial colonies on detrital materials are in a resting state rather than actively metabolizing and dividing, and they remain so until animal consumption releases the incorporated nutrients or meiofaunal movements enrich local concentrations by turbulence. In any event, bacterial incorporation of environmental nutrients increases the nutrient status of the detrital aggregate (Fig. 3–3), making aged detritus a more nutritious potential food for consumers than the originally deposited fragment of leached organic matter.

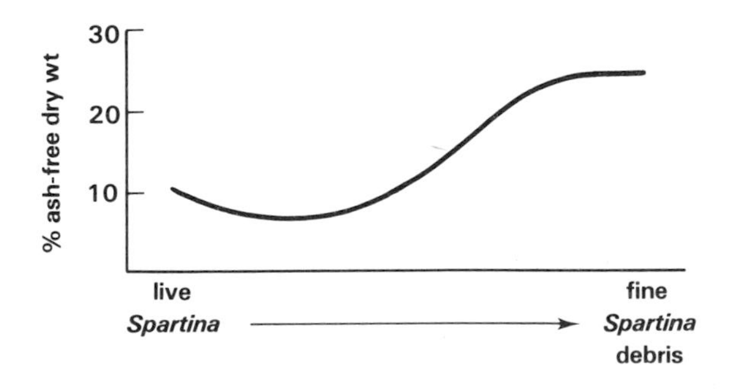

Fig. 3–3 Variation in protein content of *Spartina* debris with its state of decomposition, showing, after an initial decrease through leaching, the increase which results from microbial colonization. (After ODUM and CRUZ, 1967.)

3.2 'Detritus feeders'

3.2.1 Modes of obtaining detritus

The majority of the mud-flat macrofauna are detritus feeders in the sense that they consume living and dead organic matter associated with the substratum. Five different modes of obtaining this detritus can be distinguished.

Several sedentary polychaetes (mainly of the families Spionidae, Terebellidae and Ampharetidae) spread a number of tentacles over or through the substratum and potential food particles are collected and transported to the mouth by ciliary tracts or by peristaltic muscle contractions. Sorting of this collected material may then take place in the mouth region. Enteropneusts, for example *Saccoglossus*, feed in an essentially similar manner.

Secondly, many lamellibranchs suck detritus from the mud surface using the inhalent siphon; these in effect vacuum-clean the substratum. The body of the animal is frequently buried in the substratum and the long inhalent siphon ascends to and roves over the surface, leaving in some cases species-specific suction traces. The collected material is sieved and sorted by the gill lamellae, unsuitable material being rejected in pseudofaeces and the remainder being passed by cilia to the mouth. The gill filter is sufficiently fine in many species to collect bacteria and small algae (which may give rise to problems with respect to their subsequent consumption by man). The positioning of the body of the lamellibranch below the surface of the substratum has an obvious survival value against bird predation: in some species the body may be buried very deeply, for example a 4 cm long *Scrobicularia* may have a siphonal length of 28 cm, allowing the shell to be situated some 20 cm below the surface.

Several arthropods, for example, *Corophium* and ocypodid crabs, selectively remove suitable particles from the substratum by a mechanical sorting process, using their appendages. We will consider some of the specializations involved in this form of feeding in the next chapter.

Fourthly, a number of polychaetes consume the substratum itself, often by eating their way through it, and then digest out any nutritive material. Capitellids, maldanids and *Arenicola* ingest mud by using a protrusible fleshy proboscis and void the undigested material as a long faecal string, often on the mud-flat surface. *Arenicola*, at least, can feed using other mechanisms, for example by the capture of particulate organic matter in a sediment filter which is then consumed. *Nereis diversicolor* also possesses several feeding modes, one of which resembles the latter mechanism in the use of a filter – in this case, however, the filter is constructed of mucus.

We have mentioned that benthic algae and surface detritus may be put into suspension by the incoming tide. Suspension-feeding lamellibranchs with short inflexible siphons, for example the cockle (*Cardium*), utilize this to obtain surface material otherwise unavailable to them. Suspension-feeding lamellibranchs are, however, often rare in estuaries. This results (*a*) from the high sediment load of estuarine water, (*b*) from the comparatively limited time for feeding available to mud-flat animals which feed on matter suspended in the water column (littoral suspension feeders on rocky shores may rely heavily on wave action to increase the potential feeding time), and (*c*), in the case of animals such as mussels and oysters, because of the scarcity of hard substrata to which to attach (they are often very abundant, however, if natural or man-made hard substrata are present). The growth rate of *Cardium* on a mud or sand-flat, for example, is markedly dependent on its position on the shore: those nearest low water are much larger at the same age than those on the upper shore, as a result of their longer potential feeding time.

3.2.2 *Digestion and assimilation*

By these means, the majority of the infaunal invertebrates take into their guts the organic matter in and on the substratum in various degrees of separation from the inorganic sediment. Attempts to study the precise diets of detritus feeders have concentrated on two distinct approaches: (*i*) providing them with various presumed components of the diet in pure culture and investigating the extent to which that component is in fact assimilated; and (*ii*) studying the proportion of the total theoretically-available carbon that is assimilated on passage through the gut by comparison of the carbon content of the sediment and of the faecal materials. Relatively few studies have been able to investigate the actual diets of free-living animals because of the large number of different potential food items contained within a single detrital particle.

Some work which has looked at the effect that detritus feeders have on their environment has tended to support the hypothesis that detritus feeders preferentially select living micro-algae such as diatoms when these are available. Indeed, European species including *Hydrobia ulvae* and *Corophium volutator*, and an American gastropod *Nassarius obsoletus*, are capable of grazing down diatoms and other micro-algae very rapidly, and *Hydrobia* at least will select preferentially sediments especially rich in diatoms (COLES, 1979). Many estuarine invertebrates are now being shown to exhibit previously unthought-of powers of selective ingestion.

Experiments with detrital analogues (e.g. KOFOED, 1975) have shown that detritus feeders are capable of efficiently assimilating diatoms, bacteria and, to a lesser extent, aged detritus, whilst the 'raw' detritus itself may largely be untouched by the digestive enzymes present in the guts of most species. The basic organic debris apparently passes through the gut, being stripped on passage of its various living components (from bacteria through to meiofauna), eventually to be voided back to the environment where it serves as the substrate for further microbial colonization (as described in section 3.1.2). After an interval, during which time the faecal pellet loses its integrity and reverts to a state equivalent to that of the background sediment, and during which the microbial recolonization takes place, the debris can be reingested and the process repeated.

Relatively unselective feeders are able to digest only a small proportion of the total organic carbon present in the sediment (some 1–10%) and hence it may be concluded that the living detrital components comprise only this small percentage of the detrital carbon. Further, organic matter itself may comprise only a small percentage of the weight or volume of the sediments in or on which the detritus feeders live (e.g. Fig. 3–1). Therefore, a large quantity of sediment must be sorted and/or consumed in a given time in order to obtain sufficient assimilable food (in some cases of the order of 10 g animal^{-1} day^{-1}). By so doing, these species effect a continual reworking and mixing – bioturbation – of the surface and subsurface layers of the substratum: populations of deposit feeders may pass all the sediment surrounding them, down to a depth of some 10 cm, through their guts once every 2–5 years. Selective feeders, especially those removing diatoms, can, by virtue of their selectivity of ingestion, achieve much higher assimilation efficiencies, in some case assimilating 70–80% of the ingested carbon.

3.2.3 Density, productivity and biomass

Estuarine detritus feeders can achieve large population densities (Tables 3–1 and 3–2). Since, however, most species are small in size (Table 3–3), their biomasses are only relatively moderate (Tables 3–1 and 3–3). In part, their small individual size is a reflection of a short life span: many live for only 1–1.5 years, and several species which in the sea can attain a life-span of many years (> 10) do not achieve this in most estuaries, as seen, for example, in many cockle populations. The reduction in size of predominantly marine species within estuaries seen earlier (Fig. 2–1) is therefore at least partly due to a lowered longevity.

On mud-flats, the distribution and abundance of the various species are often patchy, each species dominating local areas or zones and the different species together forming a mosaic pattern (Table 3–2 and Fig. 3–4). The causes of this are still largely unknown. The dominant particle size of mud-flats, and with it the abundance of organic detritus (p. 24), does vary locally as a consequence of the small-scale pattern of water movement, and to some extent differential abundances may simply reflect the quantity of food present (Fig. 3–5). But interspecific interactions, including of the interference competition

Table 3–1 Indications of the density and biomass achieved by detritus feeders.

Species	*Density (nos m^{-2})*	*Biomass (g wet wt m^{-2})*
Nereis diversicolor	96 000	145
Arenicola marina	220	200
Corophium volutator	63 000	29
Hydrobia ulvae	107 800	237
Macoma balthica	56 500	75

Table 3–2 Density (nos m^{-2}) of detritus feeders at individual stations along transects of estuarine shores.

Species	*Dee (Cheshire)* (1)	*Bristol Channel* (2)	*Tamar* (3)	*Exe* (4)	*Mersey* (5)	*Dovey* (6)
Nereis diversicolor	12	2000	–	132	*c.*20	–
Corophium volutator	22 560	–	–	312*	infreq.	4 397
Hydrobia ulvae	24	2 500	14 160	5 000	infreq.	36 120
Scrobicularia plana	280	–	1 094	404	–	–
Macoma balthica	600	320	28	32	4 736	–
Mya arenaria	100	–	–	–	3 872	–

*This value refers to the related *C. arenarium*

(1) Stopford, S.C. (1951). *Journal of Animal Ecology*, **20**, 103–22.
(2) Rees, C.B. (1940). *Journal of the Marine Biological Association of the United Kingdom*, **24**, 185–99.
(3) Spooner, G.M. and Moore, H.B. (1940). *Ibid.*, **24**, 283–330.
(4) Holme, N.A. (1949). *Ibid.*, **28**, 189–237.
(5) Fraser, J.H. (1932). *Ibid.*, **18**, 69–85.
(6) Beanland, F.L. (1940). *Ibid.*, **24**, 589–611.

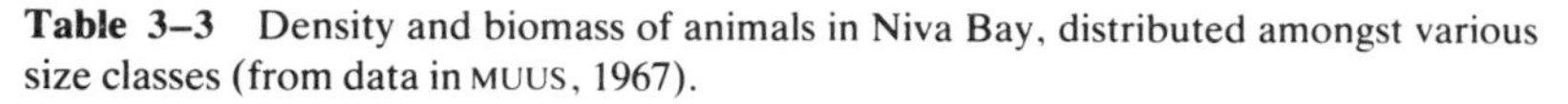

Table 3–3 Density and biomass of animals in Niva Bay, distributed amongst various size classes (from data in MUUS, 1967).

Weight range (mg)	*Density (nos m^{-2})*	*Biomass (g wet wt m^{-2})*
1 000–10 000	2	5
100– 1 000	700	95
10– 100	1 100	48
1– 10	60 000	88
0.1– 1	51 000	15
0.01– 0.1	198 000	10
0.001– 0.01	758 000	4
	1 068 802	265

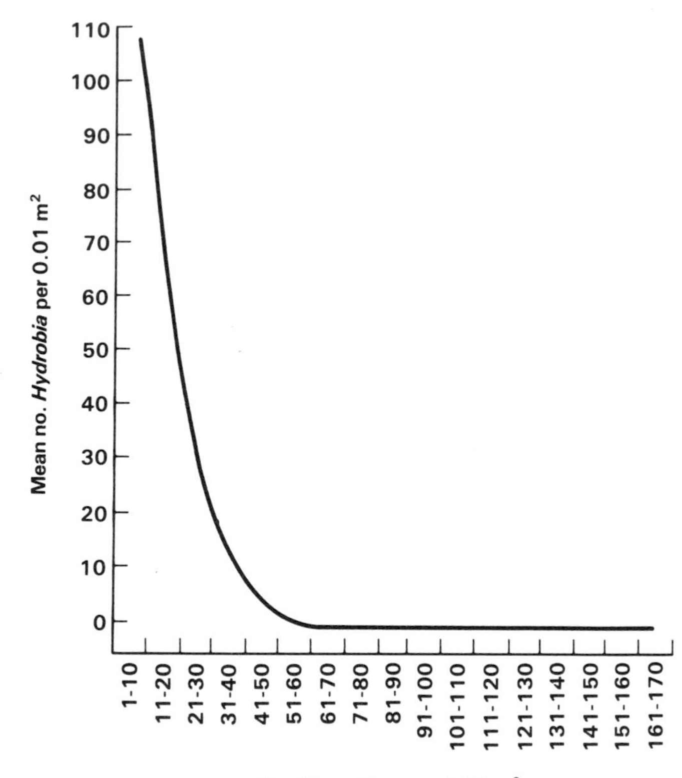

Fig. 3–4 Mutually-exclusive centres of abundance on a mud-flat: the average numbers of the gastropod *Hydrobia ulvae* found in the same 0.01 m² area as differing densities of the amphipod *Corophium arenarium*. (After BARNES and HUGHES, 1982.)

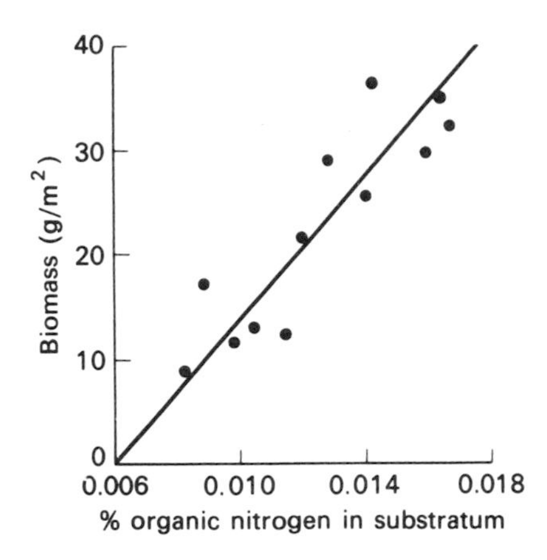

Fig. 3–5 Relationship between biomass of *Arenicola marina* and organic nitrogen in the substratum. (After LONGBOTTOM, M.R. (1970). *Journal of Experimental Marine Biology and Ecology*, **5**, 138–57.)

type, are also likely to be important. A number of species, by continually reworking the sediment, etc., are known to render the habitat unsuitable for others, whilst the feeding activity of suspension feeders and of vacuum-cleaning deposit feeders can prevent the larvae of many species from successfully settling on to the substratum surface. Such interactions can be viewed as a somewhat indiscriminate form of competition for space.

Most deposit feeders have an annual production of the same order of magnitude as their average annual biomass (WARWICK, *et al.*, 1978). Hence, considering the high primary productivities characterizing estuaries, the productivity of the primary consumers is surprisingly low. What limits their biomass and productivity? The rôle of predators as limiting factors is considered in the next section (3.3), in which it will be concluded that rarely do they appear to control the numbers of the detritus feeders. A second possibility is shortage of food. One view of estuaries (e.g. BEUKEMA, 1976) emphasizes (*a*) the abundant organic matter present in estuarine sediments, (*b*) the presence of unused detritus at considerable depth in the substratum, and (*c*) the observation that the activities of the detritus feeders do not deplete the environmental pool of organic matter. Supporters of this view, therefore, conclude that estuaries are areas of resource superabundance and that competition for food is inconceivable. A second school of thought (e.g. LEVINTON, 1972) stresses (*d*) the limited proportion of the detrital pool that is really available (i.e. can be digested) to the detritus feeders, (*e*) the manner in which the total available space, both within the mud and across the mud-flats, has been partitioned between the various species, (*f*) the grazing down of diatoms to very low levels that species such as *Hydrobia ulvae* can achieve, and (*g*) the limited bacterial productivities that result from nutrient shortage within the sediments. Highlighting these aspects leads to the conclusion that estuarine species are likely to compete for food and that they are resource and competition limited. Most of this debate is based on entirely theoretical considerations; as yet, unfortunately, there are very few critical measurements or experimental results to resolve the argument. A third and potentially important limiting factor is environmentally-induced mortality. Changing local patterns of sediment deposition and erosion can result in organisms being buried or being exposed and washed away, and tidal movement of surface-dwelling animals up the shore can cause large numbers to be stranded in a drift-line at the high spring-tide mark. Once again, however, these effects have not been adequately quantified, and so the basic question remains unanswered. We will return to this subject on p. 35.

3.3 Predators

The large densities of estuarine detritus feeders support considerable numbers and a diverse assemblage of predators. Three categories of predators may be recognized: carnivorous mud-flat invertebrates, fish and birds.

3.3.1 Invertebrates

Apart from the migratory epifaunal component (crabs, shrimps, etc.) which probably plays a dual carnivorous/scavenging rôle, the main invertebrate predators are errant polychaetes, for example *Nephtys* and some *Nereis* spp., and nemertines, for example *Lineus ruber*, which probably prey largely on small polychaetes and amphipods. The opisthobranch *Retusa alba* is one of the few predatory estuarine gastropods: it feeds to a large extent on *Hydrobia ulvae*. This component of the predator populations is generally small and has been little studied.

3.3.2 Fish

Most, although not all, of the estuarine fish are predators. One notable exception is the grey mullet (*Mugil*) which is one of the few vertebrate detritus feeders, though it will also browse algae (p. 22) and the younger stages feed on plankton. The diet of many fish is known approximately, but owing to the difficulties of observation, the quantities of the various food species taken in unit time are largely unknown. Many fish take a wide range of prey species: GREEN (1968) has recorded that at the freshwater end of the Gwendraeth Estuary, *Pomatoschistus microps* (a goby) feeds predominantly on chironomids, oligochaetes and harpacticoids, whilst in salt-marsh pools it feeds on mites, harpacticoids and *Corophium*, and near the mouth of the estuary its staple diet is barnacles. At high tide, *Pomatoschistus* feeds on *Corophium, Hydrobia, Nereis, Gammarus*, etc, from the mud-flats. Gobies are amongst the most numerous and important of the smaller predatory fish in estuaries and they themselves are consumed as a major part of the diet of larger fish such as brill.

Estuarine flatfish as a group are also catholic in their food preferences, but different species tend to concentrate on different types of animals – plaice (*Pleuronectes platessa*) feeds predominantly on polychaetes, flounder (*Platichthys flesus*) on crustaceans (e.g. mysids, *Crangon* and *Corophium*), brill (*Rhombus laevis*) on smaller fish and crustaceans, and dab (*Limanda limanda*) on the tentacular fan of *Sabella*. Flounder are present throughout the year in some estuaries, but it is mainly in the summer months that the biomass of this species increases markedly (e.g. to over 35 g (wet wt) m^{-2} in the Ythan) when more individuals migrate in to feed, and growth is rapid (Fig. 3–6). Most other flatfish are only temporary visitors to British estuaries.

Although some estuarine birds feed on the smaller fish (e.g. gulls, cormorants and herons feed on small flounder and gobies), the majority of the fish production is lost from the estuarine ecosystem, mainly to the sea.

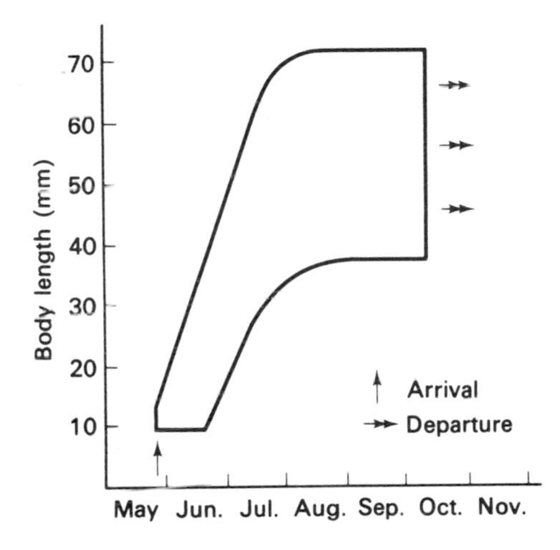

Fig. 3–6 Growth of *Platichthys flesus* in Kysing Fjord. (After MUUS, 1967.)

3.3.3 Birds

Most work on the density of predators and on the numbers of prey consumed has concerned the more easily visible avifauna. Some estuarine birds, for example many geese and some duck (Anatidae), feed exclusively on the estuarine phanerogams, whilst other duck consume both animal and plant material. However, the majority of the estuarine avifauna, including numerous waders, consume the detritus-feeding polychaetes, molluscs and crustaceans of the mud-flats.

The carnivorous Anatidae take mainly epifaunal organisms and the more shallowly-burrowing infauna (e.g. *Crangon, Corophium* and *Hydrobia*). For example, shelduck (*Tadorna tadorna*) feeds mainly on *Hydrobia* and populations of this bird, which can contain 5000 individuals, can consume

15 000 000 gastropods in a single low tide period. *Hydrobia* also forms part of the diet of a number of other duck, for example teal (*Anas crecca*) and shoveler (*A. clypeata*), but even predation rates of this magnitude would not appear to be a limiting factor to the numbers of the gastropod.

The shallowly-burrowing infauna is also consumed by the shorter-billed waders (Charadriidae, etc.), but although many of the infaunal polychaetes and lamellibranchs are hidden from bird predators in deep (i.e. 5 to 20 cm) burrows, specimens of all species come within range of the bills of larger-billed waders (Haematopodidae, Scolopacidae, etc.). These waders are mostly partial migrants, feeding on estuarine mud-flats during the winter months and breeding inland elsewhere in the summer, although in some the extent of the breeding migration is only to comparatively local sand-dunes, lagoons and moorland. During low tide periods, particularly during the daylight hours, the birds follow the ebbing and rising tide down and up the beach, probing the mud-flat with their bills. At high tide, they roost on adjacent high ground.

Most waders take a considerable variety of prey, although characteristic species are preferred depending on their local abundance, and even within a given prey species a particular size range is selected. This may not necessarily be the dominant size group of the prey population; although small specimens may be more abundant, it is usually the middle size range of prey that is consumed (e.g. by knot feeding on *Macoma*, and by oystercatcher on *Cardium*). The feeding behaviour and preferred food of various waders have been observed by a number of workers. The studies of DRINNAN (1957) on oystercatchers (*Haematopus ostralegus*), GOSS-CUSTARD (1969) on redshank (*Tringa totanus*), and PRATER (1972) on knot (*Calidris canutus*), for example, have shown that very large numbers of detritus feeders may be consumed per bird per day. These authors quote values of 315 *Cardium*, 40 000 *Corophium*, and 730 *Macoma*, respectively, for the three waders, and populations of these birds may number several thousands per mud-flat.

The effect of this predation on the prey populations varies. For some species it may be negligible, for example knot effected a maximum reduction in the numbers of *Macoma* of less than 4% per annum, and HUGHES (1970) estimated that predation by oystercatchers on *Scrobicularia*, whilst constituting the largest source of predation mortality on his studied population, nevertheless resulted in only 5–6% being consumed per year. (The deeper-burrowing individuals were beyond the reach of oystercatcher beaks, the predator never extending its bill into the mud beyond the level at which its eye neared the mud surface.) In some other studies, however, predation has been shown to be a very important source of mortality. EVANS *et al.* (1979) demonstrated that predation by waders during the winter months effected a 90% reduction in the numbers of the adults of both *Hydrobia* and *Nereis* on mud-flats in the Tees Estuary. On the Tees, however, special factors may be operating. Waders have there been confined to an ever diminishing area of mud-flat as a result of extensive reclamation of the intertidal zone (see Fig. 5–3), and this may account for the very high level of consumption in the small fragment of mud-flat remaining.

On balance, it would appear that although predation rates may sometimes be high, the migratory waders do not often control the numbers of the detritus feeders. In part this may be because feeding on relatively small individual food items is costly in time (and energy) and birds therefore concentrate on the most profitable areas of a mud-flat in which prey concentrations are particularly high, and ignore regions with a lower than threshold density; in part it may be a reflection of the limited period of the year during which waders and wildfowl are present in estuaries; and in part it may be that the populations of the birds are kept below the potential carrying capacity of their estuarine wintering grounds by mortality occurring at their breeding sites or during migration.

3.4 The food web

The terminal position of the avifauna in estuarine food chains and a highly simplified diagram of the whole food web described in this chapter are shown in Fig. 3–7. The estuarine food web can be characterized by the complex manner in which the various components are linked together, by an inefficient use of the available energy by resident species (allowing migrants to utilize some of the 'excess'), and often by either the import or the export of a large amount of fixed energy in the form of detritus. That is, the estuarine ecosystem displays a low order of maturity in the sense of MARGALEF (1963), as is also evidenced by its low biotic diversity and the relatively small community biomass supported by unit quantity of the fixed energy that is fueling the system.

The total resident estuarine biomass is relatively small because (*a*) the harsh environmental conditions may result in heavy mortality and (*b*) the typical estuarine animals are small, short-lived species of the *r*-selected type, as we saw earlier. Considerations at the ecosystem level also suggest that the community biomass is relatively small as a result of (*c*) tides and currents removing much of the energy fixed within the estuary out to the adjacent sea (especially in systems with an extensive fringing salt-marsh or mangrove-swamp component) or on to the surrounding land in the form of a strand-line (particularly where winds blow predominantly onshore and where tidal range is high) and (*d*) the severe and fluctuating physical regime ensuring that organisms have high maintenance-energy requirements for homeostasis (whether behavioural or physiological, see Chapter 4).

Various reasons for the low diversity of estuarine biotas were suggested on pp. 13–14. The unstable, unpredictable and physically highly-stressed nature of the environment provides additional causes. In order to persist in the face of a fluctuating physical environment, organisms need to be adaptable and generalist in their requirements, and it is usually considered that a given environmental resource-spectrum can accommodate many specialists but few generalists. Highly diverse ecosystems are certainly formed of highly specialist species, and they are subjected to perturbations of a lower frequency and smaller magnitude than those experienced by estuarine organisms. To inhabit estuaries successfully, the populations of macrofaunal species must have a

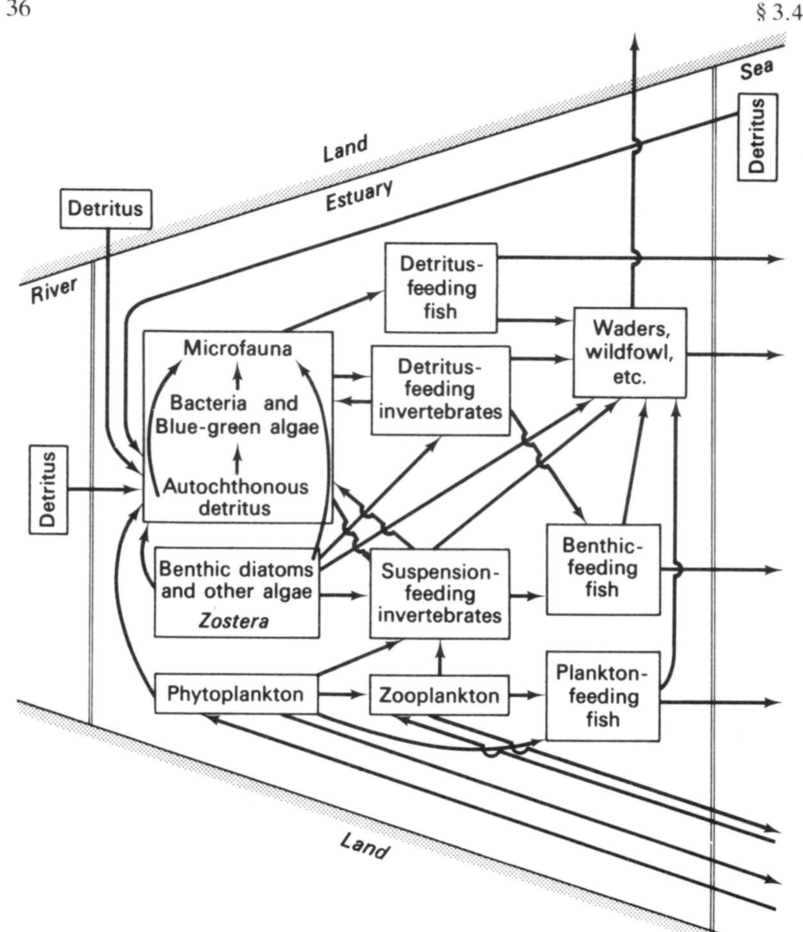

Fig. 3–7 A simplified estuarine food web.

large measure of dynamic stability in order to persist through time: such stability is found associated only with low species diversity.

Estuaries, although one of the most productive ecosystems known to man (e.g. with benthic primary productivities of up to 3.0–7.0gC fixed m^{-2} day^{-1}), therefore contrast markedly with coral reefs and rain forest, whose productivity they rival, in the manner in which their productivity is achieved. Whereas rain forest and coral reefs are productive by virtue of efficient retention, use and recycling of a limited pool of materials by a diverse, constant but dynamically unstable system; estuaries are productive because of the magnitude of the environmental through-put of effectively limitless materials, even though they are being utilized inefficiently by a low-diversity, variable but dynamically more stable community of organisms.

4 Adaptations

The physical nature of the estuarine environment is severe, mud-flat animals having to withstand both brackish water of variable salinity and the rigours of intertidal life. They must also be able to survive in the face of considerable predation. Hence, it is not surprising to find that most of the marine species which have managed successfully to colonize estuaries have evolved a number of adaptations, which it may be suggested, serve (*a*) to ameliorate the effect of the environmental conditions, (*b*) to enable the maximum quantity of potential food materials to be obtained from the inorganic sediment, and (*c*) to increase the chance of escape from predation. In this chapter, we will investigate some of these anatomical, physiological and behavioural specializations of the benthic macrofauna.

4.1 Anatomical specializations

We noted in the last chapter that the macrofaunal species of temperate estuaries have to some extent subdivided the mud-flat habitat between them, so that their local distributions take the form of a mosaic of monospecific patches, or more rarely zones. Tropical estuaries possess many more species than do temperate ones (although still fewer than in the adjacent areas of sea or fresh water), and hence one might predict that the requirement for adaptations serving to minimize interspecific competitive effects (and also predation) would be especially marked in the tropics. In this section, therefore, we will turn to consider some of the adaptations shown by the dominant crabs of tropical, subtropical and, to a lesser extent, warm-temperate estuaries.

4.1.1 Feeding

In general, the grapsid crabs of tropical estuaries possess strong jaws and feed on macroparticles – carrion, leaves, smaller crabs, etc. – whilst most of the ocypodids have weak jaws and consume organic matter from the substratum. These ocypodids have evolved an elaborate mechanism for removing the organic food from the sand or silt particles and because of the nature of this mechanism they are usually restricted to substrata of specific particle size, the actual size range preferred differing from species to species. One plausible interpretation links the evolution of such differences with the need to minimize the effects of interspecific competition. Hence, we may construct a series from sand-dwellers to those inhabiting silt. In contrast, the grapsids are wide-ranging species, which traverse many different substrata in search of food.

The tips of the chelae of most ocypodids are 'spooned' and these collect the surface layers of the substratum and pass the material to the maxillipeds. The first and second pairs of maxillipeds are modified as sorting agents and bear special setae. The outer surface of the first maxilliped possesses a thick brush of setae, whilst the inner surface of the second bears numbers of specialized 'spooned' or 'feathered' setae (Fig. 4–1). The latter are species-specific in size, shape and arrangement: in species from the sand end of the substratum spectrum, the setae terminate in large serrated cups, which decrease in size and complexity towards the silty end of the spectrum, where feathered setae are the norm.

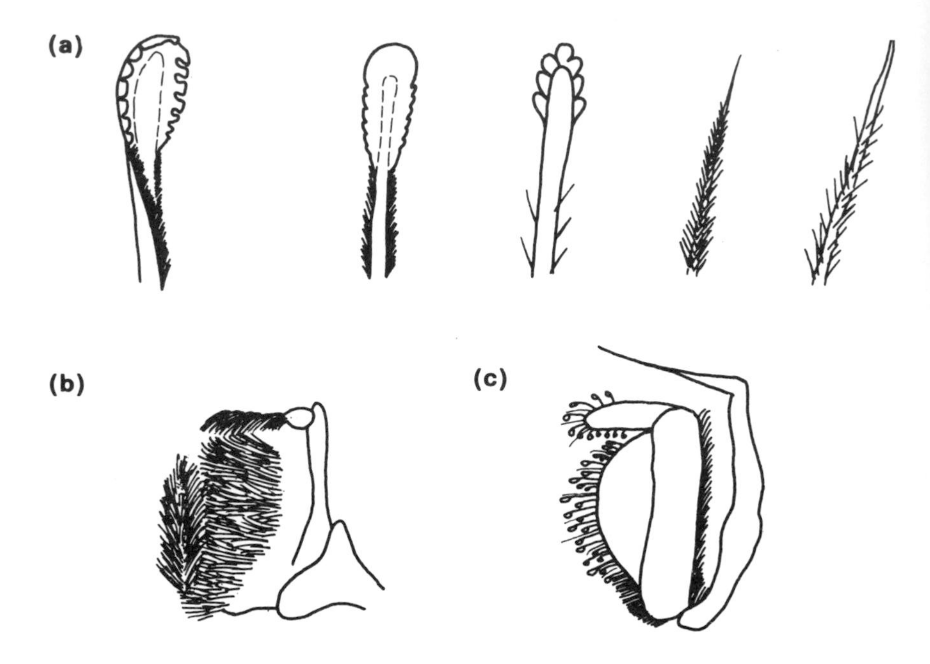

Fig. 4–1 Setal modifications in ocypodid crabs. (**a**) 'Spooned' and 'feathered' setae from various genera. (**b**) Setose brush on external face of first maxilliped of *Uca lactea*. (**c**) Spooned setae on second maxilliped of the same species (view of external face). (After ONO, 1965.)

Food material is held between the first two maxillipeds and worked between them. The spooned setae hold small sand grains against the setose brush, and diatoms, etc., on the grains are thereby brushed off. The respiratory water current is used to wash the detached organic matter towards the mouth, the respiratory intake being screened by setae to prevent the intake of food materials and sediment into the branchial cavity. This screen is of finer mesh in silt feeders than in those inhabiting sand. During this process, the heavier inorganic particles fall to the base of the buccal cavity. These then move along the edge of the third maxilliped until a ball or pellet has accumulated, this maxilliped acting as a roller plate to produce the pellet. Finally, the ball is picked off by the cheliped and deposited on the substratum.

We have seen that organic matter is less frequent in sandy than in muddy substrata (Figs 3–1 and 3–2) and so the question arises: 'How do sand-dwellers manage to obtain sufficient food materials?' Regardless of particle size of the substratum, all the species in the Tatara-Umi Estuary studied by ONO (1965) removed the same size fraction (*c*, 27 to 36 μm) from the material sorted by the maxillipeds. Since organic matter only occupies a very small percentage of the surface area of sand grains, sand-dwellers must ensure the removal of all or most of that present and must also sort a larger quantity of substratum.

We have noted above that the spooned setae reach their maximum complexity in sand-dwellers and not only are these setae more numerous, but the anatomical parts of the second maxilliped bearing them are largest in the sand species. Since a comparatively large volume of substratum must be rejected (up to 99.4%), the third maxilliped (which forms the rejected pellet) is also largest and broadest in this group. Study of the amount of nitrogen in the gut of the species investigated by Ono disclosed that all species probably obtain a similar amount of food in unit time, although there was some evidence to suggest that the sand-dwellers were the better able to assimilate the ingested material.

The differences between the species of a sand–silt series are summarized in Table 4–1. The silt-dwellers are living in an 'organic soup' and so they require fewer setal modifications. The feathered setae may, however, act in consort as a sieve to retain the smaller and lighter particles and they may also scour adsorbed organic matter from the silt particles to some extent.

4.1.2 *Gaseous exchange*

Exposure on the surface of mud-flats, on mangrove roots, etc., poses problems for relatively large animals respiring through the agency of gills. Grapsoid crabs which differ from temperate types in extending right up to the high spring-tide mark of low-latitude shores, have solved this problem by two distinct methods (VERWEY, 1930). When in air, one group, the 'pumpers', circulate the water in the branchial cavity through the gill chamber, up over the carapace and back down into the gill chamber again. The second group, the 'non-pumpers', retain the water in the gill chamber and pass a stream of air over and through this water.

In pumpers, for example *Sesarma*, *Ilyoplax* and *Macrophthalmus*, the branchial-cavity water is re-oxygenated as it passes in a thin film over the dorsal or antero-ventral surface of the carapace. This method can be very effective, *Sesarma taeniolata* being able to remain out of water for about nine hours using its pumping system. Structural modifications aid the spreading out of the water into a thin film and confine its spread so as to channel it back into the gill chamber eventually. In *Sesarma*, the area of the carapace between the afferent

Table 4–1 Feeding specializations in a series of ocypodid crabs (see text), from data in ONO, 1965.

Species	*N content* ($mg\ g^{-1}$) *in habitat substratum*	*Per cent of substratum rejected*	*3rd maxilliped :carapace proportion*	*assimilation :ingestion ratio*	*Setae*
Scopimera inflata	0.2	99.4	0.26	88.7	Large serrated 'spoons'
Uca lactea	0.4	95.8	0.12	74.4	Large 'spoons'
Ilyoplax pusilla	0.5	90.6	0.12	57.7	Narrow flattened 'spoons'
Macrophthalmus japonicus	1.0	51.2	0.07	59.8	Feathered
Paracleistostoma cristatum	2.0	11.4	0.06	–	Feathered

and efferent respiratory apertures is strongly reticulated by raised ridges and setae (Fig. 4–2), the branchial water being induced to spread over this reticulated region. In *Macrophthalmus*, granular and hairy ridges on the dorsal surface of the carapace confine the spread. Water is obviously lost by evaporation, but the crabs can take up environmental water from very thin films to replace that lost. *Sesarma* squats in small puddles and takes in water between the bases of two of the pairs of walking legs, i.e. it does not have to immerse itself to recharge. When in water, these species behave as normal aquatic crabs with respect to their respiratory water currents.

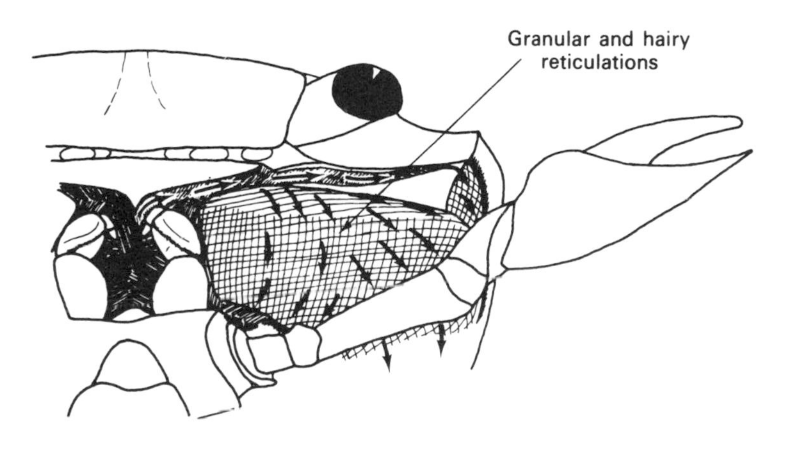

Fig. 4–2 Antero-ventral view of the carapace of *Sesarma*, showing the reticulated pterygostomial region. Arrows indicate direction of water flow (see text).

The non-pumpers, for example *Cardisoma*, *Grapsus* and *Uca*, can also take up water from thin films via an opening between the walking legs, although the position of the opening tends to be different in the two groups. The gills of these species are reduced and are thick and stiff to avoid collapse in air. The gill chamber is partitioned into an upper, vascularized chamber, functioning as a lung, and a lower, more normal gill chamber. Water can be retained within the chamber for a considerable period and replenished when convenient, although since air is continually being passed through to re-oxygenate the water, water is lost through evaporation. This is kept to a minimum by the possession of very small inhalent and exhalent apertures. The burrows of all these mud-flat crabs usually descend to the water table, so that replenishment can be effected, if necessary, in the safety of the burrow.

4.1.3 Vision

Finally, in most ocypodids the eyes are borne on the end of long ocular peduncles (Fig. 4–3), which in some species (e.g. *Macrophthalmus telescopicus*) project well beyond the lateral margins of the carapace. In life, these peduncles are held vertically upwards, with the corneas positioned close together. Elongate peduncles may confer several advantages associated with the avoidance of predators (BARNES, 1968). The most important of these is probably that the bearers can be buried below the surface of the mud, leaving only the tips of the peduncles (i.e. the corneas) at or above the mud surface. They are then effectively hidden from many of their predators (e.g. periophthalmid fish), whilst still being able to see above the surface. They are then in a position to emerge from concealment in order to feed, which they do whilst at the surface, only when it is comparatively safe so to do.

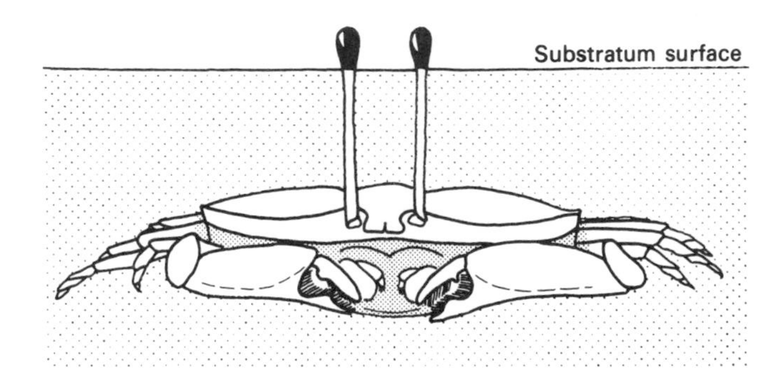

Fig. 4–3 *Macrophthalmus* buried in soft mud, with the ocular peduncles projecting through the sediment into the water or air.

4.2 Physiological specializations

4.2.1 Osmotic and ionic regulation

Pre-eminent amongst the physiological adaptations required for estuarine life are those associated with the salinity of the external medium. This field of osmotic and ionic regulation has been intensively researched by three generations of physiologists and has been exhaustively reviewed (with at least 15 major reviews in the last 20 years; for a good summary, see Schlieper, in REMANE and SCHLIEPER, 1971, and for a general statement of principles and mechanisms, see RANKIN and DAVENPORT, 1981).

Little would be gained here by attempting to summarize even the more recent items in this extensive literature, in view of the adequate and comprehensive nature of these reviews. Suffice it to point out that in most estuarine polychaetes and molluscs, the body fluids are but poorly osmoregulated, if at all, and individual body cells either tolerate wide ranges of internal concentration or to some extent regulate their own cell fluid (see below). Powers of ionic regulation are present, however; for example the potassium content of the coelomic fluid of isosmotic *Arenicola* and *Mytilus* is relatively higher than in the brackish medium. In many estuarine crustaceans, however, the permeability of the body surface is reduced and pronounced powers of osmotic and ionic regulation are possessed (Fig. 4–4), including the active uptake of environmental ions and, in some species, the production of a hypotonic urine. By these means, the body cells can be bathed by a medium of comparatively constant osmotic and ionic strength, regardless of fluctuations in the environmental salinity.

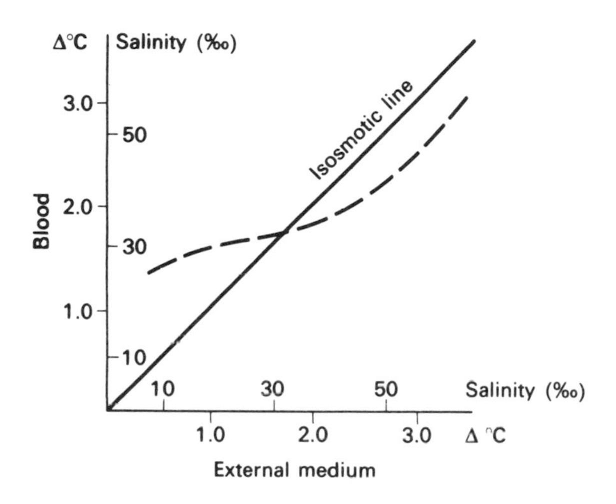

Fig. 4–4 Graph showing the osmotic concentration of the blood of *Australoplax tridentata* as a function of the external salinity (after BARNES, 1967). Δ °C = depression of freezing point.

During the last 20 years, increasing attention has been paid to the rôle of cellular amino acids in the osmotic regulation of euryhaline animals. The term 'intracellular isosmotic regulation' has been applied to the process whereby on penetration into dilute media, the concentration of free and non-essential amino acids is reduced to that level required to render the cellular and extra-cellular fluids isosmotic. On return to more concentrated media, the amino acid levels are correspondingly increased. This then reduces the osmotic gradient and may allow the cells to regulate the levels of inorganic ions more effectively. In these species, therefore, the total concentration of free amino acids is positively correlated with salinity (Fig. 4–5). Such a system has also been demonstrated in stenohaline marine species, for example echinoderms.

One must beware of the assumption that osmoregulators are 'better' estuarine organisms than osmoconformers, as it is evident that *Arenicola, Mercierella, Macoma*, etc., are as successful through the agency of wide powers of tolerance (coupled with intracellular isosmotic regulation), as are *Nereis*

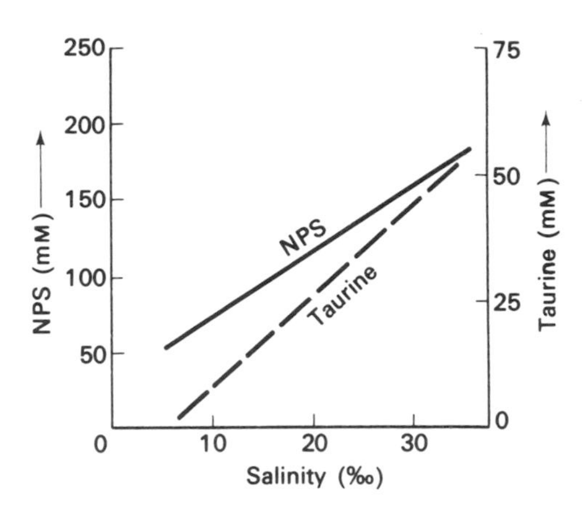

Fig. 4–5 Correlation between the external salinity and the internal concentration of taurine and ninhydrin-positive substances (NPS) in *Mytilus edulis*. (After LANGE, R. (1963). *Comparative Biochemistry and Physiology*, **10**, 173–9.)

diversicolor, *Corophium* and *Hydrobia* with comparatively feeble osmoregulatory powers, and *Palaemonetes*, *Eriocheir* and various *Gammarus* species with their more marked facilities for regulation. Some negative correlation between the extent to which a species relies on tolerance and its activity levels or behavioural complexity may be possible, however.

4.2.2 Evolutionary aspects

Most estuarine animals, and particularly those small species living in burrows or buried in the sediments, have distributions along estuarine gradients seemingly independent of salinity: they normally live well within their physiological limitations. The larger, more mobile species, however, including the crabs of both temperate and tropical estuaries, migrate up and down the salinity gradient, tidally and/or seasonally, often following the movement of specific isohalines almost exactly. Some of these appear to penetrate as far upstream as their osmoregulatory abilities will permit (BARNES, 1967).

In most temperate estuaries, the fauna comprises a series of genera each represented by only a single, or at most two, species. In a few cases, however, several species of the same or of closely related genera occur in the one estuary, and then their distributions often form a series of scarcely-overlapping segments of the salinity gradient, like beads on a string (Fig. 4–6). The distributions of some epifaunal, scavenging or phytophagous crustaceans, for example *Jaera* spp., *Sphaeroma* spp. and various gammarids, approximate this pattern, as in some areas do those of *Hydrobia* spp. It is generally assumed that the boundaries between the ranges of contiguous species are determined by competitive interactions, but what determines whereabouts within an estuary the range of a particular species will be centred?

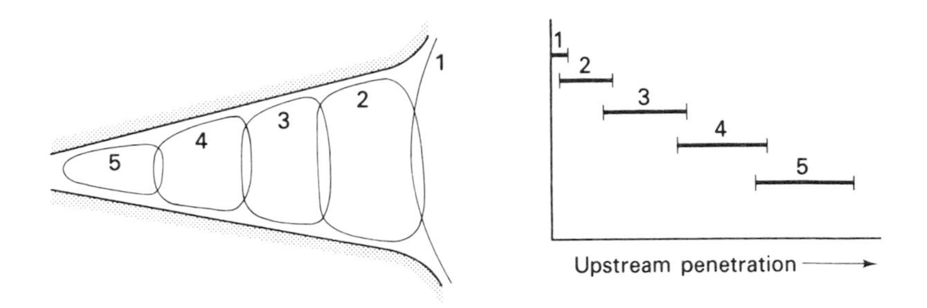

Fig. 4–6 Diagrammatic representation of the distributions of some epifaunal amphipods and isopods along estuarine salinity gradients.

Variation in morphological characters and, to a lesser extent, in behavioural reactions has been much studied, but all animals also display variation in their physiological mechanisms due to genetic differences, although this has received little attention. This is partly because it is difficult to disentangle genetic and non-genetic variation, the latter caused by sexual and size differences, by the moult and reproductive cycles, and by the effects of previous environmental conditions on the individual. When, however, these other sources of variation are 'removed', estuarine animals may still display considerable residual variation (Fig. 4–7). Populations in different geographical areas may, in isolation from each other, diverge as a result of selection acting on different sections of the pool of variation present within

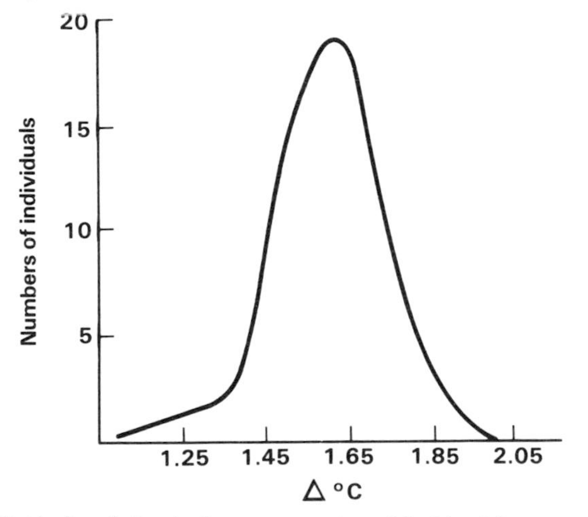

Fig. 4–7 Individual variation in the concentration of the blood (measured as freezing-point depression) within a population of the crab *Australoplax tridentata* acclimated to the same environmental salinity (20‰). (After BARNES, R.S.K. (1968). *Comparative Biochemistry and Physiology*, **27**, 447–50.)

each population (perhaps largely as a result of ecological factors specific to each area). In such a manner, geographically isolated populations may come to show different physiological reactions to salinity and to be least stressed in different parts of the overall gradient. This condition is well shown by populations of the prawn *Palaemon squilla* within the Mediterranean and Black Sea basins (Fig. 4–8).

Should these isolated populations later achieve the status of separate species, and sometime thereafter be able to expand their geographical ranges into areas occupied by other members of the species flock, then coexistence within a shared estuary would be possible if the central parts of the ranges of each species were sufficiently separated along the salinity gradient. Competition within the zones of overlap would then yield the beads-on-a-string pattern.

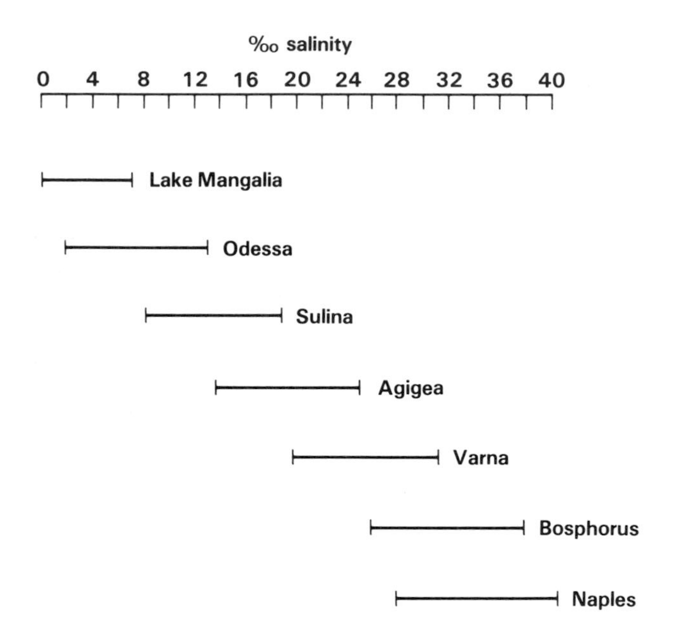

Fig. 4–8 Salinity-tolerance ranges of different populations of the prawn *Palaemon squilla* from within the Mediterranean and Black Sea basins. (From data in PORA, E.A. (1946). *Bulletin of the Institute of Oceanography, Monaco*, **903**, 1–43.)

4.3 Behavioural specializations

4.3.1 Widespread tendencies

Many behavioural adaptations are common to a large number of estuarine species: burrowing is one such. This not only allows a species to remain hidden from a number of potential predators, but, as stated earlier, it enables a species to inhabit regions of comparatively constant or only slowly changing salinity.

This considerably eases the strain on an animal's osmoregulatory mechanisms. It may also enable it to achieve a higher degree of osmoregulation, by allowing it to acclimate gradually to changes in salinity, whereas epifaunal organisms are forced to acclimate rapidly. The epifauna in general is more mobile, however, and can migrate out of regions of unfavourable salinity. Because of the low oxygen tensions of interstitial water, infaunal species may have to irrigate their burrows, and the irrigation process may also be used to obtain food. This to some extent will nullify the advantages accruing to burrowing species, as (*a*) the openings of the burrow and the respiratory water currents can advertize the animal's presence to predators, and (*b*) the irrigation water will be labile in its salinity.

Animals can also maintain a certain constancy of environmental salinity by changing their preferred position on the shore when they penetrate estuaries. Let us consider the case of a species living at about the high water level of neap tides on marine shores. If this species progressively moves its preferred position down the shore as it penetrates further into an estuary, it will remain in conditions of comparatively high ambient salinity (see Figs 1–4 and 1–5). This change of habitat is shown by many species; it has been elegantly demonstrated, for example, by SANDISON and HILL (1966) in their study of the fauna on mangrove roots in Lagos Harbour.

As several estuarine animals are substratum-specific, it would appear likely that they have powers of habitat selection, reacting to chemical and mechanical properties of the sediment, etc. That polychaete and barnacle larvae possess such powers and can delay their metamorphosis until such time as a suitable substratum is reached (within certain limits) has been known for some time. Studies on estuarine organisms have as yet concerned comparatively few species; Meadows, in a series of papers on *Corophium* (see, e.g., MEADOWS and CAMPBELL, 1972), however, has shown that species of this genus, given a choice between a number of substrata, will usually select one of characteristic particle size. Provided with no other alternatives, however, *Corophium* can burrow, successfully survive and grow in a wide range of particle sizes (it will be remembered that, in Chapter 2, *C. volutator* was listed in the shelter-specific category). Substratum selection in this species may depend upon detection of the amount of organic matter present, but further research into this aspect is required. A closely similar behaviour pattern is shown by *Hydrobia ulvae*.

Finally in this section we can consider the rhythms exhibited by the detritus-feeding macrofauna. Most are active mainly at night, which may be of advantage with respect to decreasing the chances of being taken by predators since the efficiency with which waders can find their prey is relatively low during the hours of darkness. Most are also especially active during periods of tidal cover. This is hardly surprising considering that they are essentially aquatic species, although this simple fact is easily overlooked by people able to work on estuarine mud-flats only at low tide. In *Corophium volutator*, the rhythm of swimming activity is endogenous (HOLMSTRÖM and MORGAN, 1983), but in *Hydrobia ulvae* it is exogenous, the various phases being a direct response to the presence or absence of light and water. Many arthropods of the salt-marsh surface also show endogenous rhythms which result in them moving below the surface of the substratum when the marsh is likely to be covered by flooding tides. Here an endogenous rhythm would seem particularly advantageous, since were a beetle or a mite to be caught on the surface by tidal water flow it would almost certainly be washed away and/or drown.

4.3.2 Species-specific patterns

Other behaviour or life-history patterns are rather more specific to certain types of species. We shall consider two examples: the breeding migrations of the more mobile crustaceans; and the larval and reproductive strategies of the infauna.

(i) BREEDING MIGRATIONS Although the adults of many decapod crustaceans can live successfully in low salinities, their eggs and larvae are frequently more susceptible to dilution of the medium. In a sessile species, or one with only limited powers of mobility, such a susceptible stage in its life history would bar it from inhabiting estuaries, but the majority of decapods are highly motile and hence the resistant adults can migrate into the adjacent sea for the duration of the breeding season. In some cases, the breeding migration is initiated by a change (restriction) in the adult's salinity tolerances at the onset of the reproductive season. Similar short time scale migrations, unconnected with reproduction, are also shown by many decapods. *Carcinus*, for example, migrates up and down estuaries with the state of the tide. This field has been reviewed by ALLEN (1966).

In some species, it is the egg which is the critical phase in the life history: the eggs of *Carcinus* have a minimum tolerated salinity some 24‰ higher than that of the adults. In others, for example *Palaemon longirostris*, a large prawn common in several of the Norfolk Broads, it is the larvae which are intolerant. When the larvae of this prawn are ready to hatch, the adults leave brackish water and migrate to the sea. Larval life is completed in the sea, after which period young prawns (at about one twelfth of their final length) migrate back into estuaries. Many similar examples of migrating shrimps (e.g. many *Crangon* species), prawns (e.g. *Penaeus* spp.) and crabs (e.g. *Callinectes*, *Sesarma*, etc.) could be quoted and the same phenomenon is also shown by some non-decapod crustaceans, for example several mysids; but perhaps one of the most spectacular examples is provided by the grapsid *Eriocheir sinensis*.

A native of China, this species was introduced into north-western Europe in or before 1912, probably through the agency of shipping. Since that time, it has become established from France to the Baltic countries and it has been recorded on several occasions from the Thames. It is capable of penetrating through estuaries into rivers, being particularly abundant in a number of German rivers (see, e.g., KÜHL, 1972) in which it may extend for a distance of more than 400 miles from the coast. In its native Chinese rivers it extends even further from the sea (up to about 700 miles). Yet at the approach of the breeding season, first the males and then the females migrate downstream to the estuaries, where copulation is effected; the females then continuing their migration to the sea. There the eggs develop and the larvae are released and there the whole larval life and almost the first two years of juvenile life are spent. Then, in winter or early spring, the young crabs start to migrate back through the estuaries and into the rivers. Adult females usually die after spawning, but the adult males return at the rate of some 5 to 8 miles per day; the whole return journey therefore taking up to 12 weeks. Dams and similar obstacles are surmounted by walking overland. The juveniles, of course, travel more slowly (0.5 to 2 miles per day).

(ii) LIFE-HISTORY STRATEGIES Many of the marine relatives of estuarine animals produce planktonic larval stages. As a result of the net seawards flow of water in estuaries, however, the corresponding larval types would, if produced within an estuary, be carried out into the adjacent sea. Some of the infaunal species of estuaries produce instead free-living and free-swimming larvae that do not inhabit the overlying water mass but live within the parental burrow system, as shown by *Nereis diversicolor*, or within the surface silty deposits, as in *Arenicola*. The larvae of *Hydrobia ulvae* are short-lived and although they are planktonic, they remain close to the substratum where water velocities are lowest. Some controversy surrounds the larval stages of this species and it is possible that in some populations the larvae have been suppressed. Other north-west European hydrobiids produce only a few, yolky eggs which develop, each within its own capsule, right through to the young snail stage before hatching; hence *H. ulvae* is somewhat unusual in producing many eggs which develop into free-swimming larvae.

It was mentioned above that many estuarine animals are short-lived, often annual, species. Characteristically, they breed only once and then die, i.e. they are semelparous. In semelparous species, it is usual to find that a relatively large proportion of their available energy is devoted to the production of eggs or larvae and relatively little is invested in growth. This contrasts with the pattern shown by iteroparous species (which reproduce more than once during their lives and live through several breeding seasons): these allocate a relatively large share of their resources to growth and less to reproduction. In molluscs, for example, iteroparous species on average devote some 18% of their total production to reproduction, compared with a value of 30% in semelparous forms. The only semelparous estuarine animal which has been studied in any detail in this respect is *Hydrobia ulvae*, and most surprisingly it was found to allocate only 6% of its production to reproduction. There is a possibility that some *H. ulvae* may live for more than 1½ years and manage to breed twice, but this can in no way account for such a low investment in progeny. It is not yet known to what extent other estuarine species show a similar feature.

Several species do invest energy in the provision of yolk for their eggs. An argument developed for lagoonal species (BARNES 1980), which may also apply to some extent to those of estuaries, is that brackish-water habitats often fluctuate between periods of environmentally-induced population crash and those of sufficiently large population densities as to precipitate competition for food (see p. 31). Immediately after a population crash, it would be advantageous to have an *r*-selected life history so as to pre-empt the now vacant resources with ones own descendents. At times of high population density, however, it would be advantageous to devote reproductive resources to the production of few offspring, each with its own food reserves, and so better able to cope with the high levels of competition. The life-history pattern of living short lives and producing relatively few, well-provisioned young compared to their marine relatives can on this basis be regarded as the necessary compromise. The longevity of estuarine species can be seen to be *r*-selected, permitting rapid potential increase in numbers because of the short generation time, whilst the reproductive strategies are relatively *K*-selected, allowing the young stages to withstand potential competition.

5 Estuarine Biology vs Population and Industrial Pressure

Many British estuaries are officially classified as 'grossly polluted'. The water in the Tees Estuary, for example, has been described as 'visually repulsive'; and in much of the lower part of the Mersey, 'fouling of the shore by sewage, vegetable fats, oil and other material is a prominent and exceedingly unpleasant feature' (PORTER, 1973). The problem is not confined to Britain, however: an American worker has subtitled this environment 'the septic tank of the megalopolis'; and international concern is even being expressed on the state of such large water bodies as the Mediterranean, and the seas and oceans into which estuaries discharge.

That pollution is a serious and immediate problem in many estuaries (see, e.g., ROYAL COMMISSION ON ENVIRONMENTAL POLLUTION, 1972) is basically due to the shelter and the freedom from wave action that we have previously noted in connection with the distribution of, for example, *Cardium glaucum*. Shelter from the action of wind and wave has made estuaries ideal natural harbours, and from this comparatively simple property has grown the modern concentration of population and industry to be found around many estuaries in all parts of the world. It is the discharge of effluents from these urban and industrial centres, and the reclamation of land to provide additional sites for 'development' that constitute the major sources of conflict.

5.1 Ships, people and industry

Many of the original waves of colonization of Britain and other maritime countries established settlements at the heads of estuaries. Through many centuries, these early settlements remained small, serving as trading posts around the last convenient bridging places over the rivers. In the last few hundred years, however, these small towns have been greatly enlarged, first by the development of ports, then by shipping-dependent industry, and lastly by the formation of residential suburbs. For the first two stages in this process, the safe harbour provided by the estuary remained the overriding consideration.

Increasingly, industry has tended to move to the coast. In Britain, all but one of the major areas for new investment in chemicals have coastal or estuarine sites, and oil refineries and nuclear and oil-fired power stations are similarly concentrated in this zone. In part, this is due to the saving in transport costs (for raw materials and for finished products) that results from the siting of factories adjacent to their import/export dock. Another contributory factor is the increasing size of the many industrial plants requiring cooling water: when rivers can no longer provide an adequate supply of cooling water (and are no longer able to receive large quantities of discharged waste products), industries move to estuaries.

With the movement of people towards these coastal industries and with the general increase in population, both sewage-disposal facilities and building land are becoming increasingly critical. Today, in the United Kingdom and Eire, over 30% of the population live adjacent to estuaries, and the same value also applies to many other industrialized nations, for example Japan, even including some with high 'total land area: length of coastline' ratios, such as the U.S.A. The history of the industrialization of four British estuaries has recently been summarized by PORTER (1973).

The consequences of these centres of population and industry on estuaries and on their organisms fall into four categories: introduction of alien species through the agency of shipping, in some cases aided by heated discharges; dredging of the beds of estuaries in order to maintain channels for the passage of large ships; alteration of the ecology of the fauna and flora by the accidental or deliberate discharge of pollutants; reclamation of mud-flats or salt-marshes for building purposes. Immigrant species have been mentioned above (pp. 17 and 48) and we will not consider dredging operations further, except to note that (*a*) dredging will increase the volume of an estuary, with concomitant effects on mixing processes and sediment balances; (*b*) it may affect the turbidity of the water; and (*c*) it is often associated with reclamation schemes. In the two following sections, we will investigate the two remaining categories – discharges and reclamation.

5.2 Discharges to the estuary

The variety of estuary-based industries ensures a multiplicity of discharges. The more important of these are listed below:

suspended solids, e.g. kaolin
oil
organic wastes with a high oxygen demand
hot water
toxic chemicals, e.g. cyanides and phenols
radionuclides, e.g. ^{90}Sr, ^{137}Cs and ^{106}Ru
organochlorine compounds, e.g. DDT and PCBs
heavy metals, e.g. Zn, Cd, Hg and Pb
sewage

To these we may add long-lasting or non-biodegradable 'rubbish', for example bottles, plastics and a variety of man-made and non-returnable materials, which form an artificial hard substratum in several estuaries and in the sea (even including the deep ocean trenches!).

Not all the pollutants arise from the perimeter of an estuary, however. In some, the inflowing rivers are the major contributors to the pollution load; in the case of the Humber, for example, many of the pollutants are derived from towns in the industrial Midlands and West Riding of Yorkshire.

The pollutants received by estuaries may act on organisms in several ways: they may smother; be directly toxic; be toxic to predators or herbivores after concentration by non-susceptible prey species; exert a heavy oxygen demand on the receiving water; or alter the ecology of the receiving area in a more subtle manner. Of course, any one pollutant can act in several of these modes at the same time, and different pollutants can act synergistically. In this section, we will briefly consider these modes of action: for detailed treatment of the effects of the various pollutants, the reader is referred to two symposia arranged by the Royal Society of London on the subjects of pollution in the sea (COLE, 1971) and the effects of industry on fresh waters and estuaries (RUSSELL and GILSON, 1972) and to HELLIWELL and BOSSANYI (1975).

5.2.1 *Smothering substances*

Most inert materials capable of smothering organisms and of reducing light penetration are dumped into the shallow coastal sea. Some 16 000 000 tons of colliery waste, sewage sludge, pulverized fuel ash (PFA) and various industrial wastes are tipped into the seas around England and Wales each year. The main consequences of this form of pollution in British estuaries are confined to a limited area of Cornwall, in which the china-clay industry operates.

Oil can also smother sedentary and sessile organisms, and as refineries are concentrated at estuarine sites (e.g. Milford Haven, Southampton Water, the Thames, Mersey and Humber, etc.), the dangers from accidental spills and leaks are perhaps greatest in these regions. Repeated coating by films of oil has been shown to cause the death of salt-marsh plants adjacent to an oil refinery in Southampton Water.

The effects of smothering substances can be dramatic. In 1968, HOWELL and SHELTON (1970) investigated the result of the discharge of 1 000 000 tons of china-clay waste per year into Mevagissey and St Austell Bays through the estuaries of the Par and St Austell rivers. Earlier work had shown that the fauna in polluted stretches of the Par was only a small fraction of that present in control streams, and these authors discovered 'conditions of near sterility' near the discharge points into the estuary. The smothering effect is mainly due to sediment accumulation above an organism at a rate in excess of the speed at which that organism can move upwards, although clogging of the feeding apparatus, dilution of the quantity of detritus per aliquot of substratum, reduction of light intensity, etc., may also contribute to faunal and floral impoverishment. Under natural conditions, the same effect occurs when rivers discharge large quantities of sediment into estuaries during very limited periods of time (SCHÄFER, 1972).

5.2.2 *Toxic substances*

A number of discharges into the estuarine environment contain substances capable of killing or incapacitating many organisms: cyanides and phenols rendered the water in extensive stretches of the Tees toxic to fish as early as 1931. The toxicity of a given pollutant will obviously vary with its concentration, but it may also vary quite markedly from species to species, from month to month, with the oxygen content and temperature of the water, etc., and so generalization is dangerous. Nevertheless, it is generally true that poisonous substances exert their greatest effects when the environmental conditions are most adverse for the species concerned, for example when the oxygen content is low and when the species is at the edge of its temperature or salinity range. Owing to the nature of their environment most estuarine species are 'hardy', being able to withstand considerable environmental adversity. Hence, they are to some extent pre-adapted to withstand some additives to the water, but the imposition of a further strain on their metabolism can cause death if their tolerance mechanisms are already overloaded.

The susceptibility of different animal groups to different pollutants varies widely; obviously, selective pesticides would not function without this phenomenon. Crustaceans, for example, are more sensitive to a wide variety of organochlorine compounds than are molluscs, and many fish are more sensitive to cyanides, sulphides and phenols than are most invertebrates.

In low concentrations, heavy metals, organochlorines, radionuclides and other pollutants may not be toxic to a number of invertebrates. But these substances can be concentrated within the bodies of such non-susceptible species (by factors of up to almost 1 000 000 times) to have an eventual effect on predators. The epidemic of mercury poisoning amongst the human inhabitants of Minamata Bay, Japan, is an example of this effect, the mercury having been concentrated by edible shellfish.

5.2.3 *Deoxygenation*

Perhaps the most serious form of pollution in estuaries is that resulting in a decrease in the oxygen content of the water. This condition can be achieved in three somewhat distinct ways. Some pollutants are discharged in a reduced state and are then oxidized by contact with a medium containing oxygen. Others are capable of being oxidized by bacterial action. The third category is that due to eutrophication (see below).

Very many discharges reduce the oxygen levels of estuarine water, for example sewage, cyanides, organic residues from steel, chemical and petrochemical industries, food processing plants, paper mills and distilleries. Many estuarine towns utilize the natural flushing systems of their estuaries to dispose (cheaply) of untreated sewage; for example, the crude sewage from 1 000 000 people is discharged into the Mersey, and that from a further 1 000 000 is shared between the Tees and the Humber. Much industrial waste is similarly discharged in an untreated state.

The extent to which a given discharge will deplete oxygen in the receiving area is usually measured as '5-day B.O.D.' (Biochemical Oxygen Demand), which is the quantity of oxygen used in five days for the partial oxidation of a sample of the effluent under standard conditions. For sewage, the 5-day B.O.D. represents about 70% of the total oxygen demand. With limitations, the same test can be used to determine the oxygen demand exerted by other pollutants, although for different pollutants the percentage of the total demand satisfied in the five day test period may be highly variable.

As examples of the B.O.D. exerted by all discharges into an estuary, one can quote values (from PORTER, 1973) of 32 200 kg day^{-1} for the public sewers and 220 600 kg day^{-1} for industrial discharges in the Tees, and 126 700 and 92 200 kg day^{-1} respectively in the Mersey. The total daily B.O.D. of pollutants discharged to the Tees is therefore equivalent to the removal of all the oxygen from a volume of 25 300 000 m^3 (5.6×10^9 gallons) of coastal sea water. The extent to which these oxygen demands will cause deoxygenation over a wide area will, of course, be dependent on the volume, flushing time and percentage saturation of the receiving water mass. Figure 5–1 shows two examples of the severe deoxygenation that can be caused. The results of deoxygenation of the water on the predominantly aerobic organisms in estuaries are self-evident.

To reduce the oxygen demand of crude sewage, treatment plants have been established in a number of towns, and in the Thames at least they have resulted in a marked improvement in water quality. At such sewage farms, oxidation takes place in filter beds, yielding nitrates and phosphates, etc., which are then

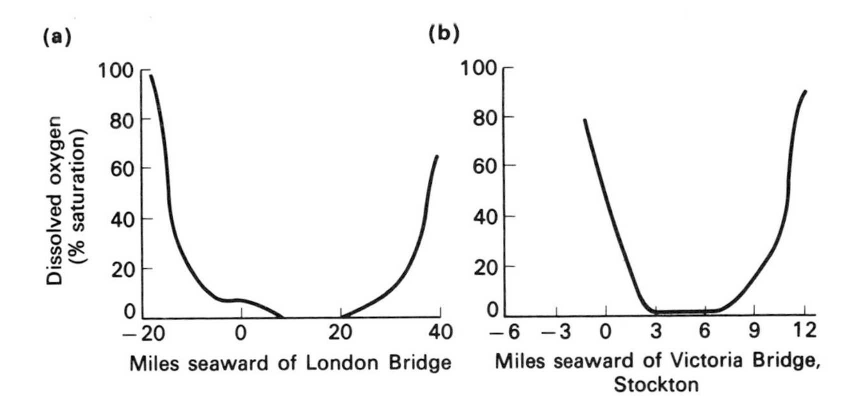

Fig. 5–1 Distribution of dissolved oxygen along two polluted estuaries. (**a**) The Thames, average value for 1950–59 compiled from measurements taken during the 3rd quarter of each year, at half tide, and under comparable conditions of fresh water flow (after BARRETT, M.J., in RUSSELL and GILSON, 1972. H.M.S.O., Crown Copyright). (**b**) The Tees (25th June 1967). (After WATSON, J.D. and D.M. (1968). *Tees-side sewerage and sewage disposal: final report*, unpublished.)

discharged to the river, estuary or sea. Although phosphates and nitrates do not exert a B.O.D., they may permit an outburst of phytoplankton production, and on the death of these plants and their subsequent decay a heavy B.O.D. may be created, causing deoxygenation of the water. Enrichment of estuaries by nutrients ('eutrophication') is not, however, a serious pollution problem in Britain, although in several areas it has encouraged excessive growth of *Ulva* and *Enteromorpha*. These green algae, although providing food for some wildfowl, blanket mud-flats and may thereby reduce the diversity and abundance of other organisms.

5.2.4 *More subtle effects*

Pollution can also change the ecology of an estuary in a less dramatic manner. The discharge of heated water does not appear to cause widespread mortalities in the British Isles. It has allowed certain immigrant species to establish themselves, however, and at the high temperatures prevailing in some enclosed water masses subject to power station discharges (e.g. Swansea Dock), these immigrants may replace native species (NAYLOR, 1965). Even in large estuaries, the discharge of warm water may enable species native to, for example, the south and west of Britain to extend their ranges northwards and eastwards (e.g. BARNES and COUGHLAN, 1973).

Thermal discharges may cause very subtle changes in an animal's ecology. Such effects on the intertidal fauna have been investigated by BARNETT (1971; 1972) near Hunterston Nuclear Generating Station at the mouth of the Clyde Estuary. This power station discharges some 91 000 m^3 of cooling water per hour (5560 gallons $second^{-1}$) into the Firth of Clyde, the discharged water being about 10°C above the ambient temperature. Barnett showed that near the Hunterston discharge *Tellina tenuis* grew more rapidly and *Urothöe brevicornis* attained a larger size than they did at Millport, some three miles away. More significantly, perhaps, he found that the breeding seasons of *Urothöe* and *Nassarius reticulatus* occurred earlier in the year near Hunterston (Fig. 5–2). Experiments suggested that, at the latter site, the eggs of *Nassarius* would also develop more quickly.

An earlier breeding season and an increased rate of egg or larval development may have serious consequences. In many littoral animals, the breeding season is timed to coincide with the period of maximum availability of food for the planktonic larvae. If the larvae and their food supply get out of phase, poor recruitment due to larval starvation is likely. American work, however, has shown that, at least in confined areas, phytoplankton blooms also occur earlier in the year when their medium is warmed by power stations. Although this will tend to restore the situation, there are additional complications. The duration of the bloom may be shortened, for example, and the relative proportions of the component species can be altered, so that planktonic larvae may still be unable to obtain an adequate quantity of the preferred food species.

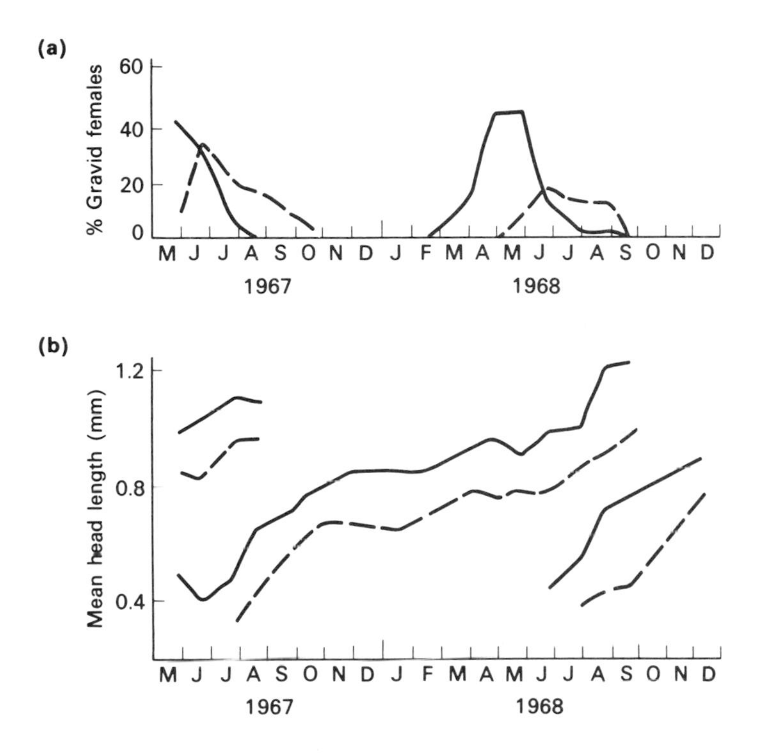

Fig. 5–2 Effects of elevated temperature on populations of *Urothoë brevicornis* subjected to power station effluent (——), compared with those away from the affected area (– – –). (**a**) Earlier onset of the breeding season; (**b**) increased mean size in three year-groups. (After BARNETT, 1971, by courtesy of the Royal Society.)

5.3 Reclamation of intertidal areas

Several large-scale operations aimed at reclaiming land from estuarine areas have been conducted, or are proposed, in countries suffering from a shortage of agricultural or building land. The silt and detritus frequent in estuaries make such reclaimed areas extremely fertile and if the land so formed is to be used for agriculture, man generally proceeds by encouraging the natural reclamation effected by salt-marsh plants. This may be achieved by planting *Spartina* on virgin mud-flats or by cutting ditches and constructing small banks on existing marshes. By these means, the high organic content of the future soil is maintained.

In contrast, the reclamation process is entirely artificial if the land is required for building purposes and if speed in reclamation is essential. A barrage is constructed at the appropriate distance offshore and material is pumped from the bed of the estuary into the enclosed region until a new land surface is formed. After settling and drying of the substratum, which may take several years, the ground may be built upon without subsidence.

The Netherlands, Britain and other European countries supply many examples of these reclamation processes (WAGRET, 1968; KNIGHTS and PHILLIPS, 1979). Since the 1930s, the Dutch have reclaimed much of the Rhine/Meuse Estuary and the Zuiderzee (now a freshwater lake, the Ijsselmeer, bordered by polders), and proposals for the reclamation of part of the Waddenzee have been put forward. More than one sixth of the present land area of the Netherlands was once estuarine! In Britain, the Tees (Fig. 5–3) and Southampton Water have been extensively reclaimed and 80 000 acres of the Wash have become land since the 17th Century. Between 1903 and 1981, salt-marshes around the Wash advanced at a rate of 12 m per year. The

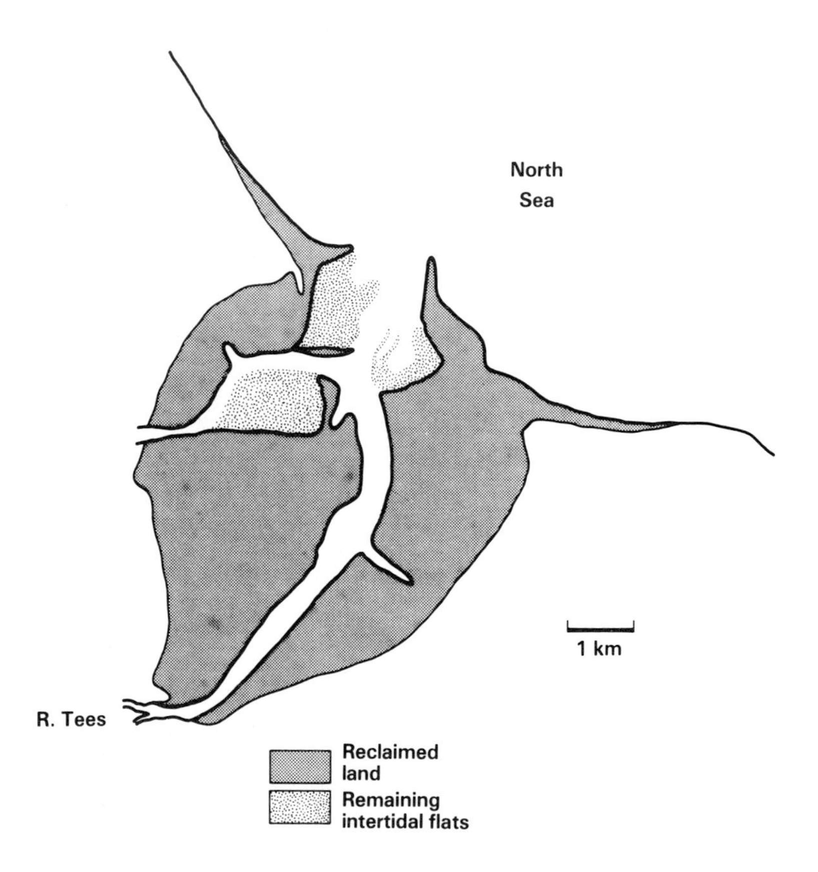

Fig. 5–3 The extent of reclamation of intertidal and subtidal regions of the Tees Estuary during the last 100 years. (After EVANS, P.R. (1981). *Verhandlungen der Ornithologischen Gesellschaft in Bayern*, **23**, 147–68.)

remainder of this century may yet see a great expansion in land reclamation in Britain; proposals exist or have been temporarily shelved for further reclamation of the Tees, for the reclamation of the Foulness flats and of 10 000 acres of the Humber for building purposes, and of parts of Morecambe Bay, the Dee, Solway Firth and further areas of the Wash for freshwater reservoir projects.

Conversion of mud-flats or salt-marshes into land or freshwater lakes destroys the habitats of many resident animals and plants, and local populations will be extinguished. As many of these organisms have wide distributions, however, this has not given rise to fears that particular species may become extinct, although continued encroachment, together with pollution on the scale of, for example, the Arno Estuary in Italy, can conceivably give rise to such possibilities in the future.

The main effects of reclamation are probably felt by the migratory estuarine birds, which use estuaries as wintering and feeding grounds and as roosts. The areas in Britain already undergoing or scheduled for possible reclamation are all particularly important sites for waders and wildfowl: Morecambe Bay supports the largest numbers of waders and the second-largest numbers of wildfowl of any British estuary; the Dee supports the second-largest numbers of waders; the Wash is the fourth most important site for wildfowl and the fifth for waders; etc. In more concrete terms, Morecambe Bay probably supports 15% of the populations of knot wintering in Europe and North Africa, whilst the Wash, Dee and Solway Firth also contribute markedly to the British total of 40% of these populations. Foulness supports 40 to 50% of the British populations of wintering *Branta bernicla bernicla*, which themselves constitute some 55% of the world population of this goose.

Removal of many of these areas will therefore have definite repercussions on the ecology of the estuarine avifauna. If only a few habitats were destroyed, the other existing sites could probably absorb the displaced individuals. But as reclamation is a general feature of much of the coastline of north-western Europe (and of the eastern seaboard of the U.S.A.), the number of suitable feeding and wintering grounds is becoming seriously limited. If this process continues, considerable diminution in numbers of several species must be expected.

5.4 Uses of estuaries

Several of the uses to which estuaries are put have been described above. Man also utilizes estuaries in other ways, however, and these may not be compatible with large pollution loads and extensive reclamation.

Although estuarine fisheries do exist, on a global scale they are insignificant and few commercially-important species are restricted to estuarine environments. Many, however, spend part of their life-cycles there. Estuaries are areas of abundant food supply, much of it unconsumed by resident species (p. 31), and this environment is therefore used by a large number of fish as a nursery area. Many animals tend to spend their adult life in mature ecosystems and their juvenile life in less mature ones (MARGALEF, 1963), so that the rapidly growing young can take advantage of the easily-available 'excess' energy in the latter: in this way, mature ecosystems exploit immature systems. Of course, the food reserves of estuaries are not only exploited by juveniles; as we have seen earlier, adults of many birds and several fish also migrate in to feed. Although, therefore, man may not catch many fish within the estuarine environment, he is nevertheless dependent on it for many of the fish which he wishes to catch in coastal seas. Land reclamation along the Dutch coast is already seen as a potential hazard to the plaice and sole fisheries (ZIJLSTRA, 1972).

By virtue of their lack of wave action and the absence of some potential predators (e.g. starfish), estuaries also form favourable areas for shellfish (lamellibranch) culture. Mussels and oysters are laid sublittorally or on rafts or ropes suspended in the water and are harvested after the appropriate growth period. In Britain, this industry is only of local importance: the value of the fishery is considerably less than £1 000 000. In other countries, however, its importance is greater. In the U.S.A., for example, the harvesting of oysters, clams and blue-crabs from Chesapeake Bay attains a yield of 5 600 000 kg $year^{-1}$ and the oysters alone were worth more than $13 000 000 in 1962.

Fishing is not only conducted to provide food; sport fishing may also earn a considerable tourist revenue. In Britain, salmon and sea-trout pass through estuaries on their spawning migrations and hence the condition of the estuarine water is critical to the continued success of this lucrative fishery. In Texas, the revenue from commercial and sport fisheries based on estuarine species is estimated at $450 000 000 per year, and comparable revenues are obtained by other American states (DOUGLAS and STROUD, 1971). Estuaries have other recreational uses, although this may appear somewhat surprising in view of the mud and the pollution described earlier. Many have yachting centres and several possess resorts near their mouths, which are dependent on tourists for their income.

Any attempt to formulate a management policy to regulate the use of estuaries must first recognize that even small modifications to the environmental regime may lead to pronounced changes in both the biology and the nature of the habitat, and that they are not self-contained systems, but help to support the productivity of the adjacent shallow coastal seas. To be successful, such policies must also reconcile the conflicting demands of, on the one hand, industry, agriculture and population pressure, and, on the other, the need to maintain high amenity values and the continued use of estuaries as nurseries and feeding grounds. That compromise is possible is almost certain, but whether it will be achieved in practice is, unfortunately, still open to debate.

6 The Scientific Study of Estuaries

6.1 History

The serious scientific study of British estuaries can only claim a history extending back about 85 years to the period when the then newly-formed Marine Biological Association of the U.K. began surveys of the estuaries of southern Devon. During the next 40 years, a series of pioneer studies of the Bristol Channel/Severn Estuary and of the Tees, Mersey, Tamar, etc., were undertaken, which laid the foundations for most subsequent work.

It is only in the last fifteen years, however, that intensive effort has been devoted to this type of habitat at several points around the coast of Britain. This period has seen the formation of (*a*) numerous regional study groups (e.g. on the estuaries of the Severn, Clyde, Forth, Tay, etc.), (*b*) a research institute, the Institute for Marine Environmental Research, specifically committed to devoting a substantial portion of its resources to the study of estuaries, and (*c*) a scientific society, the Estuarine and Brackish-water Sciences Association, to encourage the production and dissemination of scientific knowledge and understanding within this field. The latter publishes a Bulletin containing reviews of the work in progress at various estuarine sites, and is associated with the publication of a journal, *Estuarine, Coastal and Shelf Science*, the first international and multidisciplinary medium for the publication of studies on all types of brackish waters.

Shortly after the founding of this association, a number of American societies concerned with estuarine research met to inaugurate the Estuarine Research Federation in the U.S.A., which produces the journal *Estuaries*; whilst at the beginning of this period in Europe, the countries bordering the Baltic Sea set up a multinational organization under the title of the Baltic Marine Biologists to stimulate research in this area. This international activity highlights an increasing realization of the importance of estuaries to mankind and our ignorance of many of the basic biological processes occurring in brackish waters.

But this activity in no way implies that the enthusiastic amateur can no longer make any worthwhile contribution to the study of estuaries. Indeed, many estuaries remain virtually unknown to science, and many interesting and valuable data can be gathered with the aid of a pair of gum-boots, a spade and sieve, patience and inexpensive Heath Robinson apparatus. It is the purpose of this short chapter to indicate some of the more profitable lines of approach.

6.2 Field studies of distribution and abundance

We are woefully ignorant of the distribution of even the most common estuarine organisms in relation to such factors as latitude and longitude, degree of shelter, nature of substratum, amount of organic detritus, tidal level, salinity, stage in the life history, type of estuary, etc. Therefore, surveys of the littoral fauna and flora, especially in estuaries away from the larger centres of population, are still badly needed. Three techniques are required for such a survey: identification, measurement of density and biomass, and measurement of the various environmental factors.

6.2.1 Identification

All biological research requires accurate identification of the organisms under study. Hence one must first sound a warning: a misidentification is in many ways worse than no identification at all, as it can lead to misunderstanding and confusion. Let us take a hypothetical example. If one can assign a given organism to a genus, say *Nereis*, with certainty, but cannot be confident of its specific identity, a record of *Nereis* sp. or, for example, *Nereis ? diversicolor* is to be encouraged. A statement to the effect that, for example, *Nereis diversicolor* was found, on the basis that 'it was probably that species', is reprehensible.

Keys are available for several groups of estuarine organisms, but many groups remain the field of specialists. It was stated earlier that the most common macroscopic invertebrates in British estuaries are polychaetes, lamellibranchs and crustaceans. For many groups of these, the keys listed in Table 6–1 will be found useful. Most other invertebrate groups present problems; the persistent, however, may find help in the Systematics Association's booklet *Key works to the fauna and flora of the British Isles and north-western Europe* (1978, edited by Kerrich, G.J., Hawksworth, D.L. and Sims, R.W.). There are many 'popular' books available which should permit the identification of estuarine birds and fish. Flowering plants of salt-marshes, etc., may be identified by using any good 'Flora' (see Table 6–1), but other estuarine plants are less easy to name accurately.

6.2.2 Density and biomass

Although presence/absence data are useful, indications of abundance (and differential abundance in different regions) allow a much wider range of inferences to be drawn. For invertebrates, measurement of abundance requires the use of a quadrat enclosing a known area (e.g. 0.1 or 0.5 m^2) and a spade to transfer all the substratum, down to a known depth, from within the quadrat to a sieve of, for example, 0.5 or 1.0 mm mesh. After sieving, the contained organisms may be identified, counted, measured and/or weighed. Drying in an oven at 110°C for 2 days will allow dry weights to be obtained – a more reliable and accurate measure of the quantity of organic matter present than wet weight (and see below). Results can then be expressed in nos m^{-2}

(subdivided, if appropriate, into the densities of different age classes or size groups) and g (dry or wet wt) m^{-2}.

Table 6–1 Keys for the identification of common British estuarine organisms.

General

EALES, N.B. (1967). *The Littoral Fauna of the British Isles* (4th ed.). Cambridge University Press, Cambridge.

Oligochaetes

BRINKHURST, R.O. (1982). *British and other Marine and Estuarine Oligochaetes.* Cambridge University Press, Cambridge.

Polychaetes

CLARK, R.B. (1960). *The Fauna of the Clyde Sea: Polychaeta.* Scottish Marine Biological Association, Millport.

FAUVEL, P. (1923). Polychètes errantes. *Faune de France,* **5**, Lechevalier, Paris

FAUVEL, P. (1927). Polychètes sédentaires. *Faune de France,* **16**, Lechevalier, Paris.

Gastropods

GRAHAM, A. (1971). *British Prosobranch and other Operculate Gastropod Molluscs.* Academic Press, London & New York.

Lamellibranchs

TEBBLE, N. (1966). *British Bivalve Seashells.* British Museum (Natural History), London.

Isopods

NAYLOR, E. (1972). *British Marine Isopods.* Academic Press, London & New York.

Amphipods

LINCOLN, R.J. (1979). *British Marine Amphipods: Gammaridea.* British Museum (Natural History), London.

Decapods

ALLEN, J.A. (1967). *The Fauna of the Clyde Sea: Crustacea: Euphausiacea and Decapoda.* Scottish Marine Biological Association, Millport.

Salt-marsh Angiosperms

CLAPHAM, A.R., TUTIN, T.G. and WARBURG, E.F. (1962). *Flora of the British Isles* (2nd ed.). Cambridge University Press, Cambridge.

Most other groups are, or will be, covered in the *Synopses of the British Fauna* series produced jointly by the Estuarine and Brackish-water Sciences Association and the Linnean Society of London.

6.2.3 *Environmental factors*

A useful indication of many environmental factors can be obtained with relatively simple apparatus.

Sediment can be analysed into its component size fractions with a series of soil sieves (Wentworth scale) – the paper by J.F.C. Morgans (1956) in *Journal of Animal Ecology*, **25**, 367–387 will give guidance here. The quantity of organic matter in the substratum can be found by maintaining a sample at bright red heat in a crucible over a bunsen flame for 2 hours, or by incinerating the sample in an electric furnace at 700°C for 24 hours. Incineration will decompose inorganic carbonates, however, and this source of error must be rectified by flooding the sample with ammonium carbonate solution and then heating it in an oven at 110°C for 2 hours. If the sample is then weighed (to give a value of y grams) and its original dry weight (see method above) was x grams, then the weight of the contained organic matter is given by $x - y$. This is usually expressed as a percentage of the total dry weight (x).

Salinity can be calculated from the result of a volumetric titration of a water sample with silver nitrate solution ($AgNO_3$). Such a titration yields information on the 'chlorosity' of the water (i.e. the weight of chloride ions present per litre of water), but the calculations normally required to convert this into salinity (i.e. the weight of dissolved salts present per kilogram of water) can be conveniently avoided by use of the following procedure. A 10 cm^3 sample of the water to be tested is titrated with 0.16M $AgNO_3$ (using 10% potassium dichromate solution as an indicator). The volume of $AgNO_3$ used (in cm^3) then approximately equals the salinity of the water sample (in ‰). It should be noted, however, that salinity is not always the most appropriate measure of the concentration of brackish waters, since, as noted earlier, these do not approximate to standard dilutions of sea water. Hence, the concentration of individual ions is often of more interest than the total salinity. The chloride ion, as determined above, is often used as the unit of concentration of estuarine waters – in the form of the weight of chloride ions per kilogram of water, the 'chlorinity'.

Other factors can be gauged subjectively, for example the degree of shelter can be estimated from wave heights, or they can be calculated from published information. For example, although tidal heights and times can vary considerably over short distances in estuaries, if these variations are taken into account, some indication of the height of a given area on a mud-flat (in relation to the tidal cycle) may be gauged from the predicted tidal heights and tidal curves contained in the appropriate Admiralty Tide Tables. Once established from this source, contours can be plotted by the use of a surveyor's level and staff.

6.3 Observations of animal behaviour

With patience, many interesting observations can be made on a mud-flat with respect to the behaviour patterns associated with locomotion, burrowing, feeding and territorial defence. The pools ('pans') on the surface of salt-marshes are particularly good micro-environments for this form of study. If one lies quietly on the surface of a marsh and looks into a salt-pan, it is possible to observe much of the activity of *Corophium, Sphaeroma, Nereis, Hydrobia* or *Scrobicularia*, without disturbing either the animals or their habitat. Salt-pans contain few species and with a little practice the animals and their burrows can soon be recognized in the field: observe the behaviour of a given species for some time before collecting and identifying it, so that on future occasions that species can be identified on sight, without the need to collect further material.

Salt-pans are particularly interesting in that different pools may contain markedly different faunas – some will contain *Scrobicularia, Arenicola* and *Sphaeroma*, for example, whilst others may possess *Corophium, Nereis* and *Macoma*, etc. These differences do not always correlate with height of the pool above sea-level, its salinity, pH or size, etc. Careful observation of the interactions between the various species in these small pools (many are less than 2 m in diameter) may enable these differences to be explained.

6.4 Laboratory studies

Many estuarine animals survive well in captivity and they therefore make good subjects for laboratory experiments. Several estuarine species have been described as being substratum-specific (p. 15) and others prefer substrata of specific particle size when such are available (p. 47). Other macrofaunal species have the ability to detect changes in the ambient salinity, or in the concentration of one or more ions, quite accurately. Should the external conditions become unfavourable, these species may then migrate elsewhere. These behaviour patterns lend themselves well to simple experiments on 'preferences'. In essence, one provides a given species with a choice between two or more media and records which medium is preferred. Large numbers of individuals of the experimental species should be used, either together or separately, and the results subjected to statistical analysis.

Variations on such experiments are endless, but suitable apparatus for investigating (*a*) the substratum preferences of a mud-flat organism, and (*b*) the behaviour of *Carcinus* with respect to waters of differing salinity is shown in Fig. 6–1. The former test may be modified by the addition of different concentrations of a known pollutant to the sediment, in association with percentage survival counts after different time intervals.

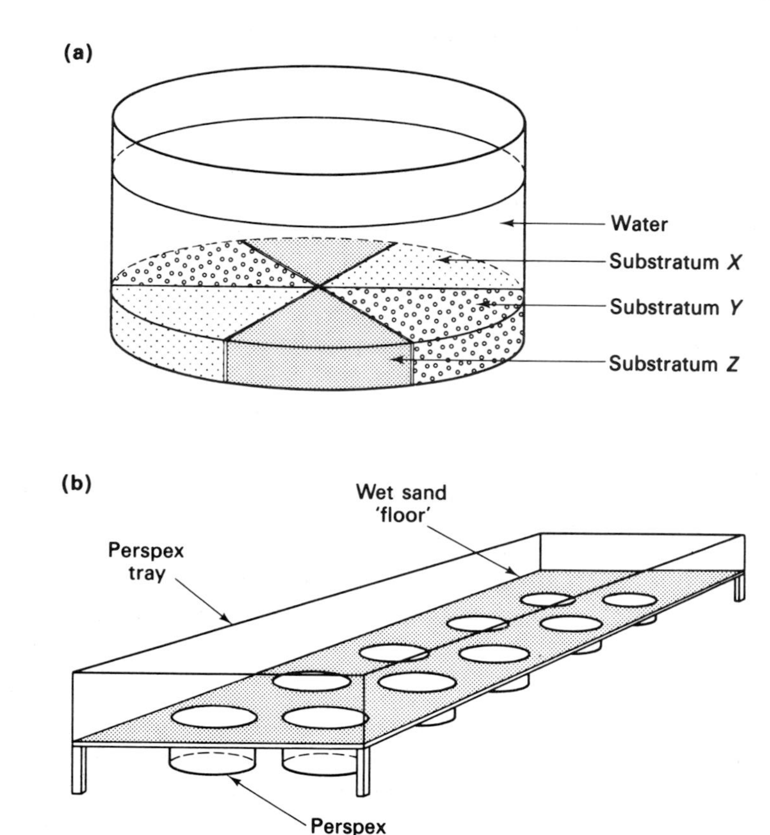

Fig. 6–1 Simple apparatus for the determination of (**a**) substratum preferences, and (**b**) the salinity preferences of epifaunal crustaceans. In (**b**), each dish is filled with water of a different salinity and should be several times larger than the experimental animal.

7 Other Brackish Waters

Estuaries are but one example from a whole range of brackish-water environments which all show certain common features (HEDGPETH, 1983). A booklet of the same nature as this one could be written on each of these other brackish habitats, but, except in one case, coastal lagoons (BARNES, 1980), such treatments are lacking, therefore brief descriptions and one or two references per environment are given below to enable the interested student to compare estuaries with other similar regions.

Besides estuaries, fjords and salt-marshes, Remane (in REMANE and SCHLIEPER, 1971) recognizes four categories of brackish waters: (1) Brackish seas, for example the isolated Caspian and Aral Seas, and the Black and Baltic Seas which are connected to the ocean by very narrow straits (KETCHUM, 1983; ZENKEVITCH, 1963); (2) coastal lagoons, i.e. bodies of water semi-isolated from the sea by sand or shingle barriers (BARNES, 1980); (3) pools on rocky shores or on salt-marshes into which sea water only occasionally penetrates (NICOL, 1935; GANNING, 1971); and (4) coastal ground-water (DELAMARE-DEBOUTTEVILLE, 1960). Various other authors also include: (5) inland saline waters, for example the Dead Sea and inland salt lakes in the U.S.A., Australia and elsewhere (WILLIAMS, 1981); and even (6) those areas of coastal seas characterized by somewhat reduced salinities (e.g. parts of the North Sea), although for all practical purposes these are better regarded as part of the marine environment. Man's activities have created a seventh category, into which can be placed several of the Norfolk Broads (ELLIS, 1965) and many coastal drainage-ditches and dykes (LAMBERT, 1930; HOWES, 1939). Such man-made environments are equivalent to small coastal lagoons and estuaries.

The environments listed above have all achieved brackish status by processes leading to the mixing of fresh and salt waters. In all but one category, the salt water is coastal in origin. The exception is the inland salt lake habitat, which has derived its salts from the substratum; consequently, ionic ratios differ considerably from that of sea water. These lakes exhibit many peculiarities, not least in their faunas which are predominantly of freshwater ancestry and which include a high proportion of insects, although salinities are often in excess of 100‰.

The source of fresh water is more variable: in rock and salt-marsh pools, it is rainfall; in coastal ground-water, it is the water-table; and in the remaining categories (lagoons, brackish seas, and estuaries), it is river flow.

Fundamentally, estuaries and lagoons differ in the width of their connections with the sea and, since the two factors are closely related, in the rate of freshwater inflow. Lagoons tend towards comparatively narrow mouths and small fresh-water inputs (p. 4). Their salinities are generally more stable than those of estuaries, the most marked fluctuations being caused by the seasonally variable evaporation/precipitation ratio. Geomorphologists distinguish between 'estuarine lagoons' receiving the discharge of rivers, and 'marine lagoons' without this form of freshwater influence – in small marine lagoons, rain water may significantly lower salinities, but it will not do so in larger examples and therefore these are outside the scope of this book. Lagoons are common throughout the world along shores subject to long-shore or onshore movement of sediment; well known groups of lagoons occur on the east coast of North America, on the west and south coasts of Africa, on the west coast of France and around much of the Mediterranean coastline, on all coasts of India and Ceylon, and in south-western and south-eastern Australia (BARNES, 1980). In Britain, they are frequent in south-west England and in East Anglia. If the barrier separating a lagoon from the influx of sea water becomes complete, contact with the sea is usually only via seepage through the barrier and such lagoons quickly become fresh, for example Slapton Ley in Devon. All lagoons are ephemeral, either reverting to the sea on destruction of the barrier, or evolving through marsh stages into land.

Most brackish seas are somewhat equivalent to large lagoons, although the isolating mechanisms in their case have been on a much larger scale than movement of sediment. The brackish seas of the U.S.S.R. have had a long and complex history. Before the Miocene, they were areas within the confines of the Tethys Sea, which some 70 000 000 years ago extended from the Atlantic, through the Mediterranean region, to what is now the Indian Ocean, with a branch connecting with the Arctic in the Kara Sea area. During the Miocene, continental movements enclosed part of the Tethys, forming the precursor of the Mediterranean in the west, and a large inland sea – the Sarmatian – to the east. The shores of the Sarmatian Sea and its derivatives fluctuated considerably during the later Tertiary, but at its major extent the Sarmatian probably included the present Black and Caspian Seas and had an extension encompassing the Aral Sea. The Black, Caspian and Aral basins later became isolated, and the Black Sea developed a connection with the Mediterranean. Therefore, these inland seas were originally relict marine enclaves, although they would have quickly become brackish. Owing to the salts brought in by rivers, ionic ratios have now departed considerably from their original relict marine state. Pleistocene climatic conditions may also have rendered some of the seas almost fresh, and so the ions derived from land drainage may now contribute markedly to the present salinities of these seas (Black, 18–21‰; Caspian, 13‰; Aral, 10‰). Their faunas contain a number of different elements, as might be expected from their histories. The relict marine or brackish species, dating back to the Tethys/Sarmatian period, are most numerous, but (*a*) freshwater species (especially fish), (*b*) a peculiar arctic component (especially in the Caspian), which possibly reached the area via fresh water during the Pleistocene, and (*c*) a Mediterranean group (particularly in the Black Sea) are also present (ZENKEVITCH, 1963). A large number of the species are endemic. The brackish seas also possess subsidiary brackish habitats (estuaries and lagoons) around their shores. By virtue of their size, habitat-diversity, and the diversity of their faunas, these inland seas can be considered the acme of present-day, brackish-water environments.

References and Further Reading

An asterisk (*) is placed against entries recommended for further reading.

ALLEN, J.A. (1966). *Annual Review of Oceanography and Marine Biology*, **4**, 247–65.

BARNES, R.S.K. (1967). *Journal of Experimental Biology*, **47**, 535–51.

BARNES, R.S.K. (1968). *Systematic Zoology*, **17**, 182–7.

*BARNES, R.S.K. (1980). *Coastal Lagoons*. Cambridge University Press, Cambridge.

BARNES, R.S.K. and COUGHLAN, J. (1973). *Essex Naturalist*, **33**, 15–31.

*BARNES, R.S.K. and HUGHES, R.N. (1982). *An Introduction to Marine Ecology*. Blackwell Scientific Publications, Oxford.

BARNETT, P.R.O. (1971). In: COLE (1971), pp. 353–64.

BARNETT, P.R.O. (1972). In: RUSSELL and GILSON (1972), pp. 497–509.

BEUKEMA, J.J. (1976). *Netherlands Journal of Sea Research*, **10**, 236–61.

BOYDEN, C.R. and LITTLE, C. (1973). *Estuarine and Coastal Marine Science*, **1**, 203–23.

BOYDEN, C.R. and RUSSELL, P.J.C. (1972). *Journal of Animal Ecology*, **41**, 719–34.

*CHAPMAN, V.J. ed. (1977). *Wet Coastal Ecosystems*. Elsevier, Amsterdam.

COLE, H.A. (1971). *Proceedings of the Royal Society of London*, **B**, **177**, 275–468.

COLES, S.M. (1979). In: JEFFERIES, R.L. and DAVY, A.J. eds (1979). *Ecological Processes in Coastal Environments*. pp. 25–42. Blackwell Scientific Publications, Oxford.

DELAMARE-DEBOUTTEVILLE, C. (1960). *Biologie des Eaux Souterraines Littorales et Continentales*. Hermann, Paris.

DOUGLAS, P.A. and STROUD, R.H. eds (1971). *A Symposium on the Biological Significance of Estuaries*. Sport Fishing Institute, Washington D.C.

DRINNAN, R.E. (1957). *Journal of Animal Ecology*, **26**, 44–69.

*DYER, K.R. (1973). *Estuaries: A Physical Introduction*. Wiley, London.

ELLIS, E.A. ed. (1965). *The Broads*. Collins, London.

EVANS, P.R., HERDSON, D.M., KNIGHTS, P.J. and PIENKOWSKI, M.W. (1979). *Oecologia, Berlin*, **41**, 183–206.

FENCHEL, T. and HARRISON, P. (1976). In: ANDERSON, J.M. and MACFADYEN, A. eds (1976). *The Role of Terrestrial and Aquatic Organisms in Decomposition Processes*. pp. 285–99. Blackwell Scientific Publications, Oxford.

GANNING, B. (1971). *Ophelia*, **9**, 51–105.

GOSS-CUSTARD, J.D. (1969). *Ibis*, **111**, 338–56.

*GRAY, J.S. (1981). *The Ecology of Marine Sediments*. Cambridge University Press, Cambridge.

*GREEN, J. (1968). *The Biology of Estuarine Animals*. Sidgwick and Jackson, London.

*HALE, W.G. (1980). *Waders*. Collins, London.

*HEDGPETH, J.W. (1983). Brackish waters, estuaries, and lagoons. In: KINNE, O. ed. (1983). *Marine Ecology. A Comprehensive Integrated Treatise on Life in Oceans and Coastal Waters*. Volume V, Part 2, 739–57. Wiley, Chichester.

*HELLIWELL, P.R. and BOSSANYI, J. (1975). *Pollution Criteria for Estuaries*. Pentech, London.

HOLMSTRÖM, W.F. and MORGAN, E. (1983). *Journal of the Marine Biological Association of the United Kingdom*, **63**, 833–50.

HOWELL, B.R. and SHELTON, R.G.J. (1970). *Journal of the Marine Biological Association of the United Kingdom*, **50**, 593–607.

HOWES, N.H. (1939). *Journal of the Linnean Society of London (Zoology)*, **40**, 383–445.

HUGHES, R.N. (1970). *Journal of Animal Ecology*, **39**, 333–56.

*JONES, N.V. and WOLFF, W.J. eds (1981). *Feeding and Survival Strategies of Estuarine Organisms*. Plenum, New York.

*KETCHUM, B.H. ed. (1983). *Estuaries and Enclosed Seas*. Elsevier, Amsterdam.

*KNIGHTS, B. and PHILLIPS, A.J. eds (1979). *Estuarine and Coastal Land Reclamation and Water Storage*. Saxon House, Farnborough.

KOFOED, L.H. (1975). *Journal of Experimental Marine Biology and Ecology*, **19**, 233–56.

KÜHL, H. (1972). *Annual Review of Oceanography and Marine Biology*, **10**, 225–309.

LAMBERT, F.J. (1930). *Proceedings of the Zoological Society of London*, **1930**, 801–8.

LEVINTON, J.S. (1972). *American Naturalist*, **106**, 472–86.

LITTLE, C. (1983). *The Colonisation of Land*. Cambridge University Press, Cambridge.

*LONG, S.P. and MASON, C.F. (1983). *Saltmarsh Ecology*. Blackie, Glasgow.

*MACNAE, W. (1966). *Advances in Marine Biology*, **6**, 73–270.

MARGALEF, R. (1963). *American Naturalist*, **97**, 357–74.

MCLUSKY, D.S. (1971). *Vie Milieu*, Supplement 22, 135–43.

MCLUSKY, D.S., TEARE, M. and PHIZACHLEA, P. (1981). *Helgoländer Wissenschaftliche Meeresuntersuchungen*, **33**, 113–21.

MEADOWS, P.S. and CAMPBELL, J.I. (1972). *Advances in Marine Biology*, **10**, 271–382.

*MUUS, B.J. (1967). *Meddelelser fra Danmarks Fiskeri- og Havundersøgelser, New Series*, **5**, 1–316.

NAYLOR, E. (1965). *Proceedings of the Zoological Society of London*, **144**, 253–68.

NICOL, E.A.T. (1935). *Journal of the Marine Biological Association of the United Kingdom,* **20**, 203–61.

NIXON, S.W. (1980). In: HAMILTON, P. and MACDONALD, K.B. eds (1980). *Estuarine and Wetland Processes with Emphasis on Modeling*. pp. 437–525. Plenum, New York.

ODUM, E.P. and CRUZ, A.A DE LA (1967). In: LAUFF, G.H. ed. (1967). *Estuaries*. pp. 383–8. American Association for the Advancement of Science, Washington D.C.

*OFFICER, C.B. (1976). *Physical Oceanography of Estuaries*. Wiley, New York.

ONO, Y. (1965). *Memoirs of the Faculty of Science, Kyushu University,* **E, 4**, 1–60.

PORTER, E. (1973). *Pollution in Four Industrialised Estuaries*. H.M.S.O., London.

PRATER, A.J. (1972). *Journal of Applied Ecology,* **9**, 179–94.

*PRATER, A.J. (1981). *Estuary Birds of Britain and Ireland*. Poyser, Calton.

PRITCHARD, D.W. (1967). In: LAUFF, G.H. ed. (1967). *Estuaries*. pp. 3–5.American Association for the Advancement of Science, Washington D.C.

RANKIN, J.C. and DAVENPORT, J.A. (1981). *Animal Osmoregulation*. Blackie, Glasgow.

*RASMUSSEN, E. (1973). *Ophelia,* **11**, 1–495.

REMANE, A. and SCHLIEPER, C. (1971). *Biology of Brackish Water*, 2nd Edition. Schweizerbart'sche, Stuttgart.

ROYAL COMMISSION ON ENVIRONMENTAL POLLUTION (1972). *Third Report: Pollution in some British Estuaries and Coastal Waters*. H.M.S.O., London.

RUSSELL, Sir FREDERICK and GILSON, H.C. (1972). *Proceedings of the Royal Society of London,* **B, 180**, 363–536.

SANDISON, E.E and HILL, M.B. (1966). *Journal of Animal Ecology,* **35**, 235–50.

*SCHÄFER, W. (1972). *Ecology and Palaeoecology of Marine Environments*. Oliver and Boyd, Edinburgh.

THORSON, G. (1957). *Memoirs of the Geological Society of America,* **67**, 461–534.

VERWEY, J. (1930). *Treubia,* **12**, 167–261.

WAGRET, P. (1968). *Polderlands*. Methuen, London.

WAISEL, Y. (1972). *Biology of Halophytes*. Academic Press, New York.

WARWICK, R.M., GEORGE, C.L. and DAVIES, J.R. (1978). *Estuarine and Coastal Marine Science,* **7**, 215–41.

WEIGERT, R.G. (1979). In: JEFFERIES, R.L. and DAVY, A.J. eds (1979). *Ecological Processes in Coastal Environments*. pp. 467–90. Blackwell Scientific Publications, Oxford.

WIEBE, W.J. and POMEROY, L.R. (1972). *Memorie dell'Istituto Italiano di Idrobiologia Dott. Marco de Marchi,* **29** (supplement), 325–51.

WILLIAMS, W.D. ed. (1981). *Salt Lakes*. Junk, The Hague.

*WOLFF, W.J. (1973). *Zoologische Verhandelingen, Leiden*, No. 126, 242 pp.

ZENKEVITCH, L. (1963). *Biology of the Seas of the U.S.S.R.* Allen and Unwin, London.
ZIJLSTRA, J.J. (1972). *Symposia of the Zoological Society of London*, **29**, 233–58.
ZOBELL, C.E. and FELTHAM, C.B. (1942). *Ecology*, **23**, 69–78.

Those interested in estuaries in relation to other coastal features, especially of a geomorphological or hydrographical nature, may find the following references useful.

BIRD, E.C.F. (1969). *Coasts*. M.I.T. Press, Massachusetts and London.
COATES, D.R. ed. (1973). *Coastal Geomorphology*. State University of New York, New York.
DAVIES, J.L. (1972). *Geographical Variation in Coastal Development*. Oliver and Boyd, Edinburgh.
DAVIS, R.A. ed. (1978). *Coastal Sedimentary Environments*. Springer, New York.
GALLOWAY, R.W. (1970). *Journal of Geology*, **78**, 603–10.
IPPEN, A.T. ed. (1966). *Estuary and Coastline Hydrodynamics*. McGraw-Hill, New York.
VAN STRAATEN, L.M.J.U. (1954). *Leidsche geologische mededelingen, Leiden*, **19**, 1–110.

Index